Getting the Most from Your Human Service Internship

Learning from Experience

Pamela Myers Kiser

Elon College

D1455961

Brooks/Cole
Thomson Learning

Australia • Canada • Mexico • Singapore • Spain • United Kingdom • United States

Counseling Editor: *Lisa Gebo*
Assistant Editor: *Julie Martinez*
Editorial Assistants: *Susan Wilson/JoAnne von Zastrow*
Marketing Manager: *Caroline Concilla*
Project Editor: *Marlene Vasilieff*

Print Buyer: *Mary Noel*
Permissions Editor: *Joohee Lee*
Cover Design: *Norman Baugher*
Compositor: *Progressive Information Technologies*
Printer: *West Group*

Printed in the United States of America
1 2 3 4 5 6 7 03 02 01 00 99

For more information, contact
Wadsworth/Thomson Learning
10 Davis Drive
Belmont, CA 94002-3098
USA
http://www.wadsworth.com

International Headquarters
Thomson Learning
International Division
290 Harbor Drive, 2nd Floor
Stamford, CT 06902-7477
USA

UK/Europe/Middle East/South Africa
Thomson Learning
Berkshire House
168-173 High Holborn
London WC1V 7AA
United Kingdom

Asia
Thomson Learning
60 Albert Street, #15-01
Albert Complex
Singapore 189969

Canada
Nelson Thomson Learning
1120 Birchmount Road
Toronto, Ontario M1K 5G4
Canada

Library of Congress Cataloging-in-Publication Data
Kiser, Pamela Myers.
  Getting the most from your human service internship : learning from experience /
Pamela Myers Kiser
    p. cm.
  Includes bibliographical references.
  ISBN 0-534-36474-8
    1. Human services—Study and teaching (Internship)—United States. I. Title.

HV11.7 .K57 1999
361'.0071'55—dc21   99-045866

 This book is printed on acid-free recycled paper.

To my family
Rick, Alison, and Emily Kiser
L.D. and Carolyn Myers

Contents

INTRODUCTION   *xiii*

Chapter 1
GETTING READY   1

So What *Is* an Internship?   1

Myths about Internships   2

Myth #1: My Fieldwork Will Probably Lead to a Job Within That Agency   3

Myth #2: While Doing My Fieldwork, I Will Learn by Doing and Will Not Have Assignments and Homework as in Traditional Classes   3

Myth # 3: My Fieldwork Is an Apprenticeship. I Will Work Under a Qualified Professional and Learn to Do as He or She Does   3

Myth # 4: As an Intern I Will Have the Responsibilities and Autonomy of a Professional Staff Person   4

Myth # 5: Doing an Internship Is Like Being a Volunteer in the Organization   4

Myth # 6: If I Do Not Enjoy My Internship, I Must Have Chosen the Wrong Field Site or the Wrong Career   4

Myth # 7: For My Fieldwork to Be Most Useful, It Should Be in the Same Type of Job I Would Like to Find After Graduation   5

Myth # 8: Everyone Gets an "A" in the Internship   5

Why Do an Internship?   6

Understanding Your Internship As Developing Over Time   7

Stage 1: Preplacement Stage   8

Stage 2: Initiation Stage   8

Stage 3: Working Stage   9

Stage 4: Termination Stage   11

Conclusion   12

Your Reflections on Getting Ready for Your Internship   13

References   13

Chapter 2
GETTING STARTED   15

Selecting a Field Site   15

Considering Populations   16

Considering Fields of Practice (Settings)   18

Considering Agency/Worker Function   20

The Initial Meeting with Your Prospective Supervisor   23

Developing a Learning Plan   26
    Identifying Goals   26
    Identifying Strategies   28
    Identifying Methods of Measurement   30

Conclusion   32
    Your Reflections on Getting Started   32

References   33

Chapter 3
GETTING ACQUAINTED   34

Getting to Know Your Agency   34

Getting to Know Your Co-workers   36

Learning About the Agency's Mission, Objectives, Goals, and Strategies   37

Learning About the Agency's Organizational Structure   40

Learning About Agency Funding   42

Learning About the Agency's Network   43

Learning About Your Client Groups   45

Learning About the Context of Your Organization   47
    Getting to Know the Community   47
    Understanding the Agency's Legal, Social, and Political Contexts   50

Conclusion   52
    Your Reflections on Getting Acquainted with Your Field Agency   53

References   54

Chapter 4
LEARNING TO LEARN FROM EXPERIENCE: THE INTEGRATIVE PROCESSING MODEL   55

The Role of Reflection in Learning from Experience   55
    Case Example   56

An Overview of the Integrative Processing Model   57

Step 1: Gathering Objective Data from Concrete Experience   59

Step 2: Reflecting   60

Step 3: Identifying Relevant Theory and Knowledge   62

Step 4: Examining Dissonance   63

Step 5: Articulating Learning   65

Step 6: Developing a Plan   67

Applying the Integrative Processing Model   69
    Case Example 1: A Direct Service Situation   69
    Case Example 2: An Administrative Services Situation   71

Your Turn to Practice Independently   75

Conclusion   80
    Your Reflections on the Integrative Processing Model   81

References   81

Chapter 5
USING SUPERVISION   82

Understanding the Supervisory Relationship   82

Supervisor Characteristics   84

Student Characteristics   86

Working Within the Supervisory Relationship   88
   Supervision Is a Planned Contact   88
   Supervision Is Prepared For   89
   Supervision Involves Discussing Professional Issues   90
   Supervision Includes Planning   91
   Supervision Generates Feedback   92

Conclusion   96
   Your Reflections on Using Supervision   97

References   98

Chapter 6
DEALING WITH DIVERSITY   99

An Overview   99

Understanding Concepts Related to Diversity   100

Experiencing Diverse Groups   101

Increasing Your Knowledge About Other Cultures   102

Developing Awareness of Your Own Culture   105

Decreasing Ethnocentrism   107
   Case Example: One Placement + Two Students = Cultural Insight   108

Understanding Biculturalism as an Adaptive Mechanism   110
   Case Example: A Client Learns to Be Bicultural   111

Understanding Agency Culture   112

Conclusion   115
   Your Reflections on Dealing with Diversity   116

References   116

Chapter 7
DEVELOPING ETHICAL COMPETENCE   118

An Overview   118

Competency 1: Understanding the Ethical and Legal Foundations of the
Profession   119
   Ethical Standards of the Human Service Profession   119
   Legal Issues and Standards of the Profession   121

Competency 2: Understanding the Values and Ideals That Should Guide Responsible
Human Services Practice   123

Competency 3: Recognizing The Ethical, Legal, and Values Issues Involved in Any
Given Practice Situation   124
   Case Example 1   125

Case Example 2   127

Case Example 3   127

Competency 4: Weighing Conflicting Ethical and Values Positions in a Given Situation and Making Decisions between Them   130

Case Example 4   131

Conclusion   136

Your Reflections on Ethical, Legal, and Values Issues   137

References   138

Chapter 8
WRITING AND REPORTING WITHIN YOUR FIELD AGENCY   139

Writing in Human Service Settings   139

Case Examples: Two Cautionary Tales on the Importance of Documentation   141

Written Reports Related to Direct Services to Clients   142

Purposes of Documentation   142

Ground Rules for Documentation   144

Common Types of Written Reports Regarding Direct Client Care   145

Written Reports Related to Administrative Services   151

Reports About Agency Programs   151

Time Interval Reports   152

Reports to the Board of Directors   152

Funding Requests   153

Public Relations   156

Oral Reporting   157

Conclusion   161

Your Reflections on Written and Oral Reporting   162

References   163

Chapter 9
MANAGING YOUR FEELINGS AND YOUR STRESS   164

Developing Self-Awareness   165

Developing Self-Understanding   166

Developing Assertiveness   167

Case Example: A Nonassertive Response to a Problem   168

Developing Conflict Resolution Skills   171

Developing Positive Self-Talk   174

Developing Self-Control   177

Learning to Manage Your Stress   179

Conclusion   184

Your Reflections on Managing Your Feelings and Your Stress   185

References   186

Chapter 10
TROUBLE-SHOOTING

187

Making Mistakes and Learning from Them   188
   Case Example 1   188

Making More Serious Mistakes   189
   Case Example 2   190

Maintaining Good Work Habits   191
   Case Example 3   192

Maintaining Appropriate Professional Boundaries   193
   Case Example 4   194
   Case Example 5   195

Guarding Against Dual Relationships   196
   Case Example 6   197
   Case Example 7   199

Maintaining a Balance between Your Work Life and Your Personal Life   200
   Case Example 8   202
   Case Example 9   205

Keeping Safe   206

Conclusion   209
   Your Reflections on Trouble-Shooting   210

References   211

Chapter 11
EVALUATING YOUR PERFORMANCE

212

Understanding the Final Evaluation   212

Preparing for Your Evaluation   215

The Field Supervisor's Perspective   218

The Evaluation Conference   219

The Faculty Liaison's Perspective   222

Conclusion   223
   Your Reflections on the Evaluation of Your Performance   224

References   225

Chapter 12
LEAVING YOUR FIELD PLACEMENT

226

An Overview   226

General Guidelines for Positive Termination   227
   Be Aware of Your Previous Experiences and Patterns with
   Terminations   227
   Be Self-Aware, Recognizing Your Needs and Wants   229
   Reflect Upon and Deal with Your Feelings   230
   Review the Experience   231
   Acknowledge the Progress and the Changes That You Have Made   233
   Anticipate and Plan for Future Needs   235

Saying Good-bye to Your Supervisor   236

Saying Good-bye to Your Clients   238

Transferring Your Work   242

Termination Rituals   244

Conclusion   246
    Your Reflections on Leaving Your Internship   246

References   247

Appendix
ETHICAL STANDARDS OF HUMAN SERVICE
PROFESSIONALS

248

INDEX

253

Preface

TO THE INSTRUCTOR

Getting the Most from Your Human Service Internship: Learning from Experience is designed for use by students and instructors who are engaged in the challenging experience of fieldwork in human service agencies. The text can be used successfully with students at any level in their educational program, by majors and nonmajors alike. Although it is written for students taking coursework in human service education programs, it may be used to great advantage with any student engaged in fieldwork in a human services agency, including those students in psychology, sociology, and other social science disciplines.

Because the structure and organization of field experiences vary broadly from program to program, the text is written with this is mind. Students will benefit from using it in course-linked fieldwork projects, one-month "miniterm" experiences, or longer, half-semester, full-semester, or full-year practica and internships. The text is ideal for use in field seminar courses where various exercises and topics can be further discussed, but it may also be used fairly independently by students.

PURPOSE

Students in their human services fieldwork are expected to meet multiple and complex objectives. The purpose of this text is to provide the information, structure, guidance, and coaching that they need in order to approach their work with greater confidence. The goal of the text is to help students maximize their learning in every experience. The text is realistic, practical, and supportive in its approach. During their fieldwork, students are supervised closely by agency supervisors but may have relatively little direct contact with faculty members and classmates as compared to their experiences in traditional classroom courses. Although fieldwork students are placed in rich learning environments and work under good supervision, the demands of the work day often preclude the opportunity for them to discuss and reflect on their experiences on a daily basis. As a result, opportunities for learning all that they might from various experiences are sometimes missed. As an instructor of field courses, I have often wished that I could sit down with each of my students at the end of the day to discuss their experiences, call attention to important issues, raise questions, and help them to draw upon the knowledge base of the profession to make sense of their experiences. Obviously, this is not possible. This text engages students in a similar process of thinking and reflection.

Another common concern among faculty teaching field courses is how to help students make meaningful connections between their classroom learning and their field experiences. This text offers specific methods and exercises that students and instructors can use to more effectively accomplish this goal. An entire chapter is devoted to teaching students a six-step model for processing their experiences, a model that includes the application of theoretical and conceptual knowledge, as well as other types of reflection.

Also, the many exercises included in each chapter take the student beyond the level of learning content and into critical thinking and active application of content to their own experiences.

PHILOSOPHY AND APPROACH TO EXPERIENTIAL LEARNING

The text reflects certain assumptions about experiential education, about the role of writing in thinking, and about how students learn. The text assumes that, while experiential education can be a powerful pedagogical approach, students often need to *learn how to learn* most effectively from experience. Although experience can be a good teacher, at times it may not teach very effectively in and of itself. Without careful thought, analysis, and reflection, students can easily draw erroneous conclusions and make incorrect inferences based on their experiences, just as they can misread or misinterpret a text. Experience is a powerful teacher, that proves to be more effective, when combined carefully with critical thinking, self-evaluation, and reflection. This text is designed to keep students engaged in this thinking process throughout their fieldwork.

Furthermore, the text requires students "to think on paper." This feature is based on the premise that writing helps the thinker to think more clearly and precisely. In each chapter, students express their thoughts in writing in response to specific exercises and prompts as well as in more open-ended reflection. As students' thoughts are expressed in writing and shared with the instructor, a fruitful dialogue can be initiated. Through reacting to the student's written work, the instructor has regular opportunities to provide supportive assistance, corrective feedback, prompts toward further reflection, or other responses that can enhance the student's learning.

Finally, the text is based on a philosophy of active, student-centered learning. Each chapter actively engages students by consistently bringing their experiences and thoughts into the discussion along with theoretical and academic content. Reinforcing the importance of the student's voice in the discussion, boxes are included in each chapter that carry excerpts from previous students' writing related to the topics discussed. This inclusion of student work enlivens the text by offering concrete examples of student experiences and reactions.

CONTENT

Much of the text's first three chapters deal with practical matters in beginning the field experience, focusing on issues such as myths about internships, stages in the development of internships (Chapter 1), development of a learning agreement (Chapter 2), and getting to know the field placement agency (Chapter 3). Chapter 4 deals with helping the student understand more about the processes involved in learning from experience and introduces a six-step model that students can use in thinking through their experiences in the field. The model is my original work, building upon earlier models of experiential learning, and is a useful framework for helping students to extract maximum learning from their experiences. In summary, this model calls upon students to be careful observers of their experiences, to reflect upon personal reactions, to identify and apply relevant knowledge, to identify dissonance (i.e., points of discomfort or conflict which might include ethical dilemmas, conflicts between theoretical points of view, etc.), to articulate their learning from the experience, and to make plans for the next step in their work and in their learning. This process not only helps students learn more during their fieldwork but also teaches them a method for thinking through their experiences that can serve them well throughout their careers.

Chapters 5 through 9 deal with special skills of human services professionals. These include learning to use supervision (Chapter 5), dealing with diversity (Chapter 6), developing ethical competence (Chapter 7), preparing oral and written reports (Chapter 8), and dealing with emotions and stress in the workplace (Chapter 9). Chapter 10, "Trouble-Shooting," serves to raise students' awareness of some potential problems in human services internships and to help students avoid these problems. The chapter covers such topics as maintaining personal and professional boundaries, guarding against dual relationships, and maintaining personal safety. The final chapters of the text move toward providing closure for the student as Chapter 11 focuses on the final evaluation and Chapter 12 deals with leaving the internship.

In some cases, the chapters have been ordered in a way that reflects the stages of the internship's development. Chapters 1 through 3, for example, focus on getting a good start in the internship, while Chapters 11 and 12 focus on bringing it to a satisfactory close. Other chapters, such as Chapter 6, "Dealing with Diversity," and Chapter 7, "Developing Ethical Competence," are less clear in terms of their logical placement in this order. In making decisions about how to order the chapters for their maximum effectiveness, instructors should give thought to the objectives of the specific field experience in which students are engaged and the emerging readiness of students to examine various issues. For example, Chapter 10, "Trouble-Shooting," though placed relatively late in the text, might be assigned earlier in many situations due to its focus on preventing some common problems and pitfalls in fieldwork.

Throughout this book, content is included in the form of case studies and examples of various kinds. This material was inspired by nearly 20 years of experience in working with students and clients in various contexts. In no case, however, does this book include information drawn from any particular student or client. The people with whom I have worked over the years have provided a rich history upon which to draw, but their struggles and situations appear in this book in very disguised, composite forms. Names, identities, situations, and details have been routinely altered to such an extent that no actual person or persons are depicted. Any likeness to the names and circumstances of real people is strictly coincidental.

ACKNOWLEDGMENTS

I wish to acknowledge gratefully the contributions of many people who contributed to the successful creation of this work, including the following: my husband and children (Rick, Alision, and Emily), who were unfailingly patient, encouraging, and supportive as I spent many long hours at the computer; faculty colleagues and administrators at Elon College, who not only provided interest in and enthusiasm for my work, but also a semester's sabbatical to begin developing the manuscript; my students at Elon College, who have taught me much of what I know about experiential education; my colleagues in the Southern Organization for Human Service Education and the National Organization for Human Service Education who encouraged my efforts, especially Ed Neukrug, who suggested that I contact Lisa Gebo, editor at Wadsworth-Brooks/Cole, about my work; and Lisa Gebo, whose constant support and guidance were indispensable throughout the writing, revision, and production process. Finally, I wish to acknowledge the time and contributions of the following reviewers who made astute and insightful suggestions, adding significantly to the quality of the finished work: Alisabeth Buck, Tacoma Community College; Mary Di Giovanni, Northern Essex Community

College; John Hancock, Fitchburg State College; Jeffrey Haber, Metropolitan State College of Denver; Paul Hand, Anna Maria College; Ed Neukrug, Old Dominion University; Theresa A. Bowman Downing, Thomas Edison State College; Mary Kay Kreider, St. Louis Community College–Meramec; Lynn McKinney, University of Rhode Island.

Introduction

One of the best ways to learn about human services and prepare yourself for a human service career is through fieldwork—that is, actually spending time in a human service agency engaged in its daily work. Most graduates of human services programs report that their field experiences were among the most important, valuable, and enjoyable parts of their professional preparation. As a student in a human service program, you have probably participated in other types of experiential education as well because learning from experience has always been heavily emphasized in human services education. Since human services education began in the 1970s, students have participated in simulations and case study analysis, group discussions have revolved around student field experiences, and extensive fieldwork has been required for degree completion. *The National Standards for Human Service Worker Education and Training* (Council for Standards in Human Service Education and Training [CSHSE], 1995) defines field experience as "a process of experiential learning which integrates the knowledge, skill, and attitudes that are concurrently being taught in the classroom" (CSHSE, 1995. p.18). Further, these standards assert that programs "shall provide each student field experience that is integrated with the rest of the training and education" (CSHSE, 1995, p. 18).

Human service programs vary in the structure of their field component. *The National Standards for Human Service Worker Education and Training* states, "While there is agreement that field experience is a critical component of any human service training program, there are variations in format, duration, and placement of the field experience" (CSHSE, 1995, p. 18). These standards suggest that field experience "should begin early in the program and be integrated with classroom" work (CSHSE, 1995, p. 18). There are indeed many formats for human service field programs, but you will probably find that your department's field program includes some combination of the following components:

1. Students engage in fieldwork early in their academic program while enrolled in one or more related academic courses. This type of experience is often referred to as a *field practicum* or a *service-learning project.*
2. Students engage in fieldwork at approximately the midpoint in the program, after having studied human services content in traditional academic courses. Field experience at this point is especially useful as it enables students to apply their previous theoretical learning to their fieldwork and then return to coursework with greater understanding and insight based on direct experience in the field.

3. Students engage in fieldwork at or near the end of their academic program. This experience, often referred to as an *internship,* serves as a capstone experience, allowing students the opportunity to apply and test what they have learned in the classroom as well as an opportunity to gain new knowledge and skills.

Despite the fact that human services students are nurtured and developed within such a tradition of active, experiential learning throughout their course of study, you may find that you are approaching your fieldwork with a sense of anxiety, a fear of not being sufficiently prepared. As one student said, "What if I go into my internship and find out I don't know anything? What if I fail?" Although this anxiety is an understandable and normal part of starting a new experience, in most cases there is little basis for this fear in reality. The foregoing discussion illustrates that students entering a field experience at any point during their human service program are not cast into the practice world without knowledge, information, or preparation for the task at hand.

Perhaps the greatest challenge for students in a human service field experience is not one of *possessing* knowledge but one of *making use of* that knowledge in practical ways. The field component should be central to your learning, serving as an integrative experience in which you retrieve your previous learning and apply it to practice situations. This objective can be quite daunting, especially in view of the broad curriculum that you have probably studied within your human services program. Content in human services education covers a wide range of topics and skills, including, for example, information about special populations and human problems; theories of human behavior and human development; skills in working with individuals, families, groups, and organizations; professional ethics; multicultural awareness and skills; and development of self-awareness. Despite its challenges, your task during your fieldwork is to make use of this material (and more), applying it accurately and skillfully to practice situations that you will encounter in the field. Through your field experience, your academic learning can come alive and take on new meaning as you see the connections between the knowledge and skills you have gained in the classroom and your "real-world" practice experiences while working in a human services agency. This dynamic connection between the academic and the practical makes fieldwork experiences in human services especially satisfying, interesting, and challenging.

With all of this in mind, the goals of this text are (1) to help you integrate theoretical and conceptual information with your experiences in the field, (2) to help you learn more from your experiences in the field by thinking extensively and carefully about those experiences, and (3) to provide the information, structure, and coaching necessary for you to explore the relatively unfamiliar territory of a fieldwork experience with confidence. Toward these ends, you will find that the text includes useful information about every stage of the process from preparing for the experience (*Chapter 1,* "Getting Ready") to ending the experience (*Chapter 12,* "Leaving Your Internship"). Each chapter calls upon you to be an active learner, reacting to, applying, and reflecting upon the many ideas discussed. As you work through the material and exercises in the text, they will guide and support you through the various stages of your field experience, helping you to seize its opportunities, anticipate and avoid its potential pitfalls, and extract maximum learning from your experiences.

The human service literature uses a variety of labels to refer to field experiences in human services. The terms *practicum, fieldwork,* and *internship* are frequently used to denote various types of field experiences. Nomenclature varies from program to program and has been the subject of much discussion in human services programs in recent years (Simon, 1999). For the sake of clarity, the terms *fieldwork, field experience,* and *internship* are the terms generally used in this book, and they are used interchangeably.

Now as we embark on your fieldwork experience together, best wishes to you! Keep in mind what a privilege it is to be allowed the status of an "insider" within a professional organization. Be determined to use every opportunity to advance your learning and development as well as to serve others. These attitudes will set the stage for an unforgettable and invaluable learning experience.

REFERENCES

Council on Standards for Human Service Education and Training. (1995). *National standards for human service worker education and training programs.* Fitchburg, MA: Author.

Simon, E. (1999). Field practicum: Standards, criteria, supervision, and evaluation. In H. Harris & D. Maloney (Eds.), *Human services: Contemporary issues and trends* (2nd. ed., pp. 79–96). Boston: Allyn & Bacon.

Getting Ready

If this is your first fieldwork experience, you probably have many questions about it. You might feel uncertain about what to expect as well as unclear about what is expected of you. As a result of this ambiguity you might find that you have some mixed emotions about doing an internship, ranging from anxiety and dread to excitement and eagerness. This chapter will help you to feel more comfortable as you approach your internship by giving you a clearer idea of what an internship is (and isn't) and the types of learning goals it is meant to accomplish. In order to bring the internship into clearer focus, the chapter considers some common myths about internships as well as faculty and student perspectives on why internships are valuable.

A Student's Reflections on Preparing for Internship

I can't wait to do my internship! I have been looking forward to this for two years. I must admit that I do have some worries and lots of questions about the internship though. For example, how do I go about getting an internship placement? Once I have a placement, how am I supposed to fit an internship into my life? I don't have time for my family and friends even now! What will be expected of me? Will I be like an employee? a volunteer? How much will I work? All day? Every day? Will I work the same hours each day? It seems that I have more questions than answers at this point, but I am looking forward to it anyway. One thing I do know is that I will get some real life experience in the field that I hope to work in after college. I appreciate the opportunity to do that.

SO WHAT IS AN INTERNSHIP?

It can be difficult to form a clear and cohesive picture of what an internship will be like, especially if you have never participated in such an experience before. One of the factors that contributes to this difficulty is the fact that internships can take many forms. For example, in some programs, the time that students spend in the internship is like a full-time job. Students work full days every day in their placements, and their curriculum is arranged so that no other coursework interrupts the day. Such internships are generally referred to as "block placements." Other programs use "concurrent placements," meaning that students perform their internships while also being enrolled in traditional classroom courses. In such arrangements, students attend their placements on alternating days or establish other types of part-time schedule arrangements.

What makes an internship worthy of the name is not how the time is arranged or whether it is part-time or full-time, but the total amount of time spent in the placement, how that time is used, and the degree of reflection, application of previous learning, and other forms of critical thinking that are included in the experience. A number of definitions of the term *internship* offered in the literature touch upon such issues. The Council on Standards for Human Services Education uses the terms *fieldwork* and *internship* synonymously and defines these terms as:

> "the advanced or culminating agency-based experience which occurs toward the end of a student college/university experience. This usually requires close supervision from agency personnel and the college/university faculty including regularly scheduled seminars with a faculty member. The experience of fieldwork provides a bridge between the academic experience and later professional employment" (1995, p. 21).

Cantor echoes the idea that the internship serves to link the academic experience with the professional world when he says, "Internships are experiential learning activities that bridge academia and the profession for students" (1995, p. 71). The same theme is touched upon in the following definition, which also reflects something of the multifaceted nature of internships as well:

> Part audition, part test, part question, part answer. Lasting for a summer or, during the school year, for a semester, interns are usually college students 'trying out' their career by working in a professional (or near-professional) capacity (Fry, 1988, p. 7).

The following description, though too lengthy to be a definition, is probably a clear enough statement to be helpful and yet broad enough to be inclusive of most field experiences in human services: The human service internship is an experience in which a student, sponsored by an educational institution, engages in education and training while working in a human service organization or role for a substantive period of time. The student engages in an ongoing process of observation, practice, and reflection in order to learn from experience. From so doing, the student gains new knowledge and skills, applies academic knowledge to practical situations, refines previously acquired skills, and gains a greater understanding of self and the human services profession.

A Student's Reflections on the Internship

At my internship I am treated as a regular staff member and not as an intern most of the time. In fact, I sometimes forget that I am still a college student. It's great to be treated as an equal and to know that my opinions are valued by others. I have the same responsibilities as the staff when it comes to dealing with the residents. As an intern, though, I have been given a broader experience so that I can learn more. I not only work with the residents but with their families as well, so part of my day is spent with the family workers. The staff is letting me see the whole operation and not just one aspect. Best of all, I am gaining confidence every day, and I'm learning things about myself that I never knew before.

MYTHS ABOUT INTERNSHIPS

Even with clearer definitions of what an internship is, students can still enter the experience with some unrealistic expectations. Under such circumstances, it is inevitable that students will encounter problems, disappointments, and misunderstandings. Therefore, an essential step in getting ready for your internship is clearing up some possible myths and misconceptions that students might hold about internships. As with most situations in life, you will experience far more satisfaction and less frustration if you understand beforehand what is realistic and unrealistic to expect in the situation.

Myth #1: My Fieldwork Will Probably Lead to a Job Within That Agency

Although some internships do extend directly into employment for the student in the host organization, this is more the exception than the rule. Nevertheless, it is important to recognize that your experience there can, and probably will, make you far more employable. Not only will your skills be enhanced by your experience but your resume, your references, and the content of your job interview responses will improve in quality as a result of your field experience. Also internships provide numerous networking opportunities in which you can meet professionals from other agencies within the human service system. These contacts can be a great asset when you conduct your search for a position. It is fair to expect that your field experience will help you secure a job, but in most programs relatively few, very fortunate students are actually employed by their agency immediately following the field experience.

Myth #2: While Doing My Fieldwork, I Will Learn by Doing and Will Not Have Assignments and Homework as in Traditional Classes

Although you will no doubt learn by doing, this is not the only manner in which learning takes place during your fieldwork. Assignments vary from program to program and from professor to professor, but typically students are required to write and read extensively during their field placements. Students are generally required to write reflection papers and/or daily journal entries about their experiences, research papers about the problems and populations with which they are working, and/or book and article reviews based on a reading list developed for the internship. Additionally, students attend class regularly (often weekly), meeting with other students in a seminar-style setting to discuss their experiences. Final grades are usually based on the quality of students' work on academic assignments as well as the quality of their work in the agency.

Myth #3: My Fieldwork Is an Apprenticeship. I Will Work Under a Qualified Professional and Learn to Do as He or She Does

An apprentice is generally thought of as an individual who is learning technical skills from a more experienced and skilled person. Although it is true that you will be working under an experienced supervisor and will have the opportunity to observe that supervisor and others as potential role models, the learning involved is more complex than the above statement implies. Human services work is sometimes described as both an art and a science (Gladding, 1992; Young, 1998). This refers to the fact that although some tasks may follow a clear protocol of how to proceed, many tasks in human services rely to some extent upon personal style and judgment as to how and when they are done. Your field experience will require that you observe, analyze, critique, and evaluate yourself and others. Out of this effort your own personal style of helping people and working with others will emerge. Although you will probably emulate various role models as you develop your own professional style, the critical thinking and independence required of a developing human service professional go somewhat beyond that of an apprentice.

Myth #4: As an Intern I Will Have the Responsibilities and Autonomy of a Professional Staff Person

After years of sitting in classrooms, you are probably eager to be active and highly involved in your agency. It is sometimes frustrating to find that as the experience begins you are asked to observe others, read related materials, and observe meetings in which you may be able to participate very little. This period of time is a predictable stage in the development of the internship. You need time to become acclimated to the agency and learn basic procedures and policies. This period of time also allows your supervisor and other staff members to become familiar with you and to become comfortable with your capabilities. As the placement progresses over time, students are typically allowed more autonomy. The degree of responsibility and autonomy that a given student is allowed varies considerably depending upon the nature of the agency, liability and legal constraints within the agency, the student's level of personal and professional skills, and the field instructor's supervisory style. Students should not expect to assume the role of a professional automatically but should recognize that, depending upon the particular situation, this status may be earned as the field experience progresses.

Myth #5: Doing an Internship Is Like Being a Volunteer in the Organization

Since most students have been volunteers before but have not been interns, they sometimes enter their fieldwork agencies assuming that there are similar expectations. Like volunteers, most students are not paid for their work during the field experience, but the similarity between volunteer work and fieldwork ends here. Volunteers are generally asked to do whatever the agency needs to have done, whether this is direct work with clients, clerical tasks, or even building and grounds maintenance. In contrast, the central focus of a fieldwork experience is the student's learning rather than the agency's needs. The activities of a student should be identified and planned based primarily on their educational value. In most field placements, as in most jobs, the student will perform a variety of tasks, some with greater educational value than others. A quality field experience, however, will engage the majority of the student's time in activities that develop and refine professional knowledge and skills. If you should find that the majority of your time is being spent in tasks with low educational value, you should discuss this with your field supervisor and with your faculty liaison.

A further distinction between a volunteer and an intern is that a greater level of responsibility is generally expected of an intern. Since interns are preparing for professional careers in human services, they are expected to adhere to professional standards in their work. Professional dress, reliability, promptness and other characteristics of professionals are clearly expected of interns. Volunteers, on the other hand, may be granted more latitude in such matters. As distinct from volunteers, interns generally enjoy certain privileges as well. Most interns, for example, have access to client records and participate in such professional-level activities as staff meetings, case conferences, and in-service training sessions.

Myth #6: If I Do Not Enjoy My Internship, I Must Have Chosen the Wrong Field Site or the Wrong Career

Just as with classroom courses, some of the most valuable field experiences are those that push you outside your comfort zone. This means that it is necessary to be uncomfortable at times to experience the greatest personal and professional growth.

Enjoyment should not be the measure of your satisfaction with your placement. Rather, it is more useful to examine how much you are being challenged, both personally and professionally, by the environment and experiences that it has to offer.

Myth #7: For My Fieldwork to Be Most Useful, It Should Be in the Same Type of Job I Would Like to Find After Graduation

Although this statement seems to be common sense, it is not always true. While your career goal is important to keep in view, it is often not possible or even advisable to enter a field placement in this exact position. There are usually a number of developmental steps to be taken to achieve professional competence in a given role. Your fieldwork can often best be used to help you achieve one or even several of these developmental steps. For example, a student who was interested in becoming a family counselor was disappointed to find that she could not secure a placement in this field due to concerns of privacy and confidentiality as well as her level of training. In her human service program, three field experiences were required. For these three opportunities she was placed in a day-care center, a battered women's shelter, and a group home for emotionally and behaviorally disturbed children. Although none of these placements offered her the exact role to which she aspired, each of them contributed knowledge and skills that were important to her overall preparation for the role of family counselor.

Myth #8: Everyone Gets an "A" in the Internship

Grades in fieldwork courses often follow a distribution more similar to classroom courses than many students expect. As discussed under Myth #2, a number of different types of assignments are factored into the internship grade. There is often wide variance in student performance levels in the field sites, ranging from students who perform exceptionally well to those who perform adequately to those who perform below average or even unsatisfactorily. Before arriving for the first day of your internship, you would be well advised to read the course syllabus carefully in order to understand the various assignments and requirements of your fieldwork experience and to understand how the different components will be weighted in assigning your final grade. Also, you should pay special attention to the items on the evaluation instrument that your field supervisor will use to evaluate your performance. Becoming familiar with these documents will clarify the exact expectations and standards that you need to meet in order to earn a grade that you can be proud of.

Exercise 1.1

What is your reaction to the myths discussed above? Which of the myths, if any, did you hold before reading this section? What new insights have you gained from this discussion?

WHY DO AN INTERNSHIP?

The discussion in this chapter has set the stage for your thinking more clearly about your own internship and the expectations you have for it. Now that you understand the general parameters of the field experience, it is time to bring your own internship into clearer focus. The following exercise asks you to think about and describe your own hopes and expectations for this experience.

Exercise 1.2

Before embarking on any journey, it is wise to clarify exactly why you want to make the trip. Why do you want to do an internship? What do you hope to gain from this experience? Take a moment to think about this and record your thoughts.

Each person has a different set of experiences, circumstances, and expectations influencing why he or she wants to do an internship. As a result, responses to the question "Why do fieldwork?" will vary broadly. Because academic programs in human services invariably require field experiences, faculty members must also answer this question. What rationale do faculty members put forth to support the fieldwork requirement? Some common rationales are summarized next. Examine these and compare them to your own earlier responses.

- Fieldwork helps to identify and modify student attitudes that interfere with effective practice (Dore, Epstein, & Herrerias, 1992).
- Fieldwork helps to develop certain personality traits in the student that are conducive to more effective practice (Dore, Epstein, & Herrerias, 1992).
- Classroom learning, even with extensive use of simulations and case studies, is not sufficient to produce effective human services professionals. Students need an opportunity to practice their skills in the real world.
- Fieldwork allows students an opportunity to apply and test classroom learning in practical, concrete situations. Students can learn to develop and test hypotheses based on theories, concepts, and knowledge that they have gained in the classroom.
- Professionalism can best be taught in the field as students have an opportunity to observe professional models over a significant period of time and think critically about their work. Students also learn to evaluate their own work and set goals for further development as professionals.
- Students can learn more about a particular subfield within human services that they might wish to enter. This direct experience can help students dispel false impressions they might hold about the field as well as gain additional skills or knowledge needed to work effectively in that field.
- Fieldwork allows students to integrate and synthesize learning that has been fragmented in the academic curriculum. Knowledge, skills, attitudes, and values must

be unified and used in conjunction with one another rather than be treated as distinct entities.

- Students need an opportunity to make a transition from the academic world to the working world. The habits formed within the academic environment are sometimes not those conducive to success in the working world.
- Fieldwork encourages personal growth in ways that differ from the classroom. The active learning, risk taking, decision-making, and personal responsibility that are inherent in agency-based practice are conducive to growth and developing professionalism.

Exercise 1.3

As you reflect on your reasons for participating in fieldwork alongside the rationales frequently cited by faculty members, what observations do you make? In what ways are your ideas similar to and/or different from those of faculty members?

UNDERSTANDING YOUR INTERNSHIP AS DEVELOPING OVER TIME

Although you might think about your internship as one, unified experience, you will more likely find that it develops and changes over time, somewhat like a living organism. Nothing is born fully developed, including your internship, so it is helpful and realistic to expect that your fieldwork experience will change as time goes by. Many models have been suggested for understanding and describing the development of internships. For example, research on music therapy internships (Grant & McCarty, 1990) and psychology internships (Lamb, Barker, Jennings, & Yarris, 1982) reveal evidence for a five-stage model, consisting of preentry preparation, finding a place in the organization, gaining identity and role differentiation, emergence of professional competence and independence, and separating. Inkster and Ross (1998) describe a similar six-stage model of internships: arranging the internship, orientation, reconciling expectations with reality, productivity and independence, closure, and reentry. Sweitzer and King (1999) present a five-stage developmental model for the internship that includes anticipation, disillusionment, confrontation, competence, and culmination.

Some writers have suggested that human services internships progress through stages similar to those of the helping process itself, including such stages as exploration, relationship building, and goal setting (Grossman, Levine-Jordano, & Shearer, 1991; Siporin, 1982), while others have suggested stages of internship development that parallel the eight stages of human development described by Erik Erikson, including stages such as developing trust, establishing autonomy, and developing initiative (Kerson, 1994). A number of writers have described the development of internships in a far more general way, as a series of tasks and issues that shift in focus as the internship moves from beginning, to middle, to end (Chiaferi & Griffin, 1997; Michelsen, 1994; Pettes, 1979).

Clearly, there is much agreement that internships do progress through certain developmental stages but little agreement as to the exact nature of these stages. The four stages of internship development discussed in the subsequent section of this chapter reflect a synthesis of ideas from a number of different models in conjunction with my own observations from working with numerous interns. Understanding some of the predictable ways in which your internship is likely to develop and change over time will better prepare you to master the challenges of each stage.

Stage 1: Preplacement Stage

The first stage of the internhip's development occurs before you arrive for the first day of work. This first stage is critical to the success of the placement because important foundations for your work are laid during this time. Usually the student and the faculty members of the human services department work together to decide upon an appropriate field placement. Efforts are made by you and/or faculty members to secure a placement compatible with your learning needs. Once a positive contact has been made with an organization, you will make an initial visit for an interview with your field supervisor. Both parties, you and your potential supervisor, form initial impressions of one another during this time. Your supervisor will be particularly interested in forming an initial impression of your interests and learning needs. If you both see the placement as appropriate and workable, the agency begins to make preparations for your fieldwork. Your field supervisor will plan for your arrival by giving thought to such issues as the types of assignments to give you and the location of a work space for you. Colleagues, and in some cases even clients, will be informed and prepared for your impending arrival. As you approach the internship, you too will be making necessary preparations. Clearing your schedule of any responsibilities that could interfere with your internship, engaging in any reading that might prepare you for your responsibilities, as well as more mundane activities, such as pulling together an appropriate wardrobe, are all essential parts of your personal preparation. (The discussion and exercises in Chapter 2, "Getting Started," will help you to work through the steps involved in this stage of the internship successfully.)

Stage 2: Initiation Stage

The first days and weeks of an internship might best be described as an orientation and a series of multiple beginnings. As you meet new people and become acquainted with the agency, you will probably experience a range of emotions, including excitement, anxiety, and anticipation. With this emotional intensity can come a burst of energy and eagerness to dig in and get started on the work of learning by doing. In contrast, as discussed earlier, supervisors and agency administrators may take a more measured approach to starting the internship. There is often reading and observation that they wish you to do before taking on any direct responsibilities. In the initiation stage your supervisor will also try to assess more fully your strengths and limitations in order to make appropriate assignments for you. Although students vary in their readiness for the internship and working independently, many express a need for structure and direction during the initiation stage of the internship. An orientation to the agency, a structured schedule of meetings that you might attend, and a set of tasks that you are to accomplish are particularly helpful during this time. If you find that your needs for structure and focus are not being met, tactfully approach your supervisor to discuss this.

Just as your internship will develop and change over time, your relationship with your supervisor will do so as well. During the early days and weeks of your relationship, a critical task is developing mutual trust. Direct, open communication is essential to building this trust. The process of establishing your learning agreement (discussed in detail in Chapter 2) is one of the most important tasks of the initiation stage and is an excellent vehicle for establishing open communication with your supervisor about your learning needs in the internship. Also essential in building trust is proving your own trustworthiness by being reliable, prompt, and eager to assume responsibility.

Although each stage of the internship offers unique opportunities for you to grow and develop your skills, each stage can offer potential pitfalls as well. A brief discussion here of these pitfalls will better equip you to avoid them. One common pitfall of the initiation stage is a tendency to "rush to judgment" about the quality of a placement. This judgment may be positive, launching you into a honeymoon phase in which you believe you have the most perfect internship and supervisor that any student could ever have. This point of view may be wonderful while it lasts, but all too often ends painfully later when this idealized view of the situation is marred by the realities of everyday demands in the organization. The student who is overly optimistic about the internship might not be sufficiently careful and assertive in establishing a clear learning agreement, feeling unduly confident that the experience is going to be ideal. As a result, the learning contract might not be clear enough or provide sufficient challenge to ensure that the student has a quality experience throughout the internship.

Of course, a negative rush to judgment is equally problematic, if not more so. Negative judgments can understandably result when, for example, a student enters the organization on a particularly chaotic day, is treated rudely by staff, or is ignored. At times student expectations of an organization and that organization's reality can be so far apart that major adjustments are required of the student. If you should find this to be the case in your internship, you will no doubt feel both frustrated and disappointed. In such a situation, however, try to keep your attention focused on two central questions, "What can I learn here?" and "How can I learn it?" All organizations have their strengths and weaknesses. Most students have much that they can learn, even in less-than-ideal settings. If your field site has some problems, changing this situation may be largely outside of your control. What is within your control is your own level of motivation and determination to learn. In the end, these characteristics will be at least as important to your learning as are the qualities and characteristics of the setting in which you are placed.

In a similar vein, Chiaferi and Griffin point out that "all too often, students focus on what they deem to be the shortcomings or personality defects of their supervisors while failing to recognize their own contribution to difficult interactions" (1997, p. 26). This observation suggests that students would do well to work on their own self-awareness during this initiation stage. Although it is inevitable and appropriate that you evaluate the quality of your agency and your supervisor, it is even more important to evaluate your own behavior and attitudes continuously, reflecting upon how they are affecting your ability to establish a sound foundation for your internship.

Stage 3: Working Stage

As relationships are developed, the learning agreement completed, and expectations voiced and negotiated, the internship begins to move into a third stage, which might be referred to as the working stage. During this stage less energy is invested in establishing relationships and negotiating mutual expectations and more energy is invested in achieving your learning goals and in accomplishing the agency's work. The trust that has been developed between you and your supervisor by this stage generally results in your supervisor allowing you more autonomy and allocating to you more responsibility. Also, by this time in the internship, both you and your supervisor have achieved a better understanding of your strengths and your learning needs. In fact, an early signal that your internship is entering this stage may be that you find yourself gaining comfort in asking questions and admitting gaps in your knowledge. This important shift enables deeper, more meaningful learning to occur.

Another characteristic of the working stage of the internship is that daily activities settle into more of a routine. You have acclimated to the agency culture and learned agency procedures. As a result, earlier anxieties are reduced and you may generally feel more relaxed and confident. This increased comfort and confidence, in conjunction with a greater understanding of the learning opportunities available within the organization and your supervisor's increased knowledge of your learning needs, might result in some renegotiation of your learning agreement at about midterm in your internship (Pettes, 1979). At this stage in the experience, both you and your field supervisor are in a better position to plan and implement learning experiences specific to your particular needs.

In the working stage of the internship, the student engages deeply in the process of learning to be a professional. Student learning during this stage focuses on such issues as developing practice skills, learning about agency administration, recording and other forms of written communication, developing professional responsibility, learning to use supervision, and learning to juggle multiple tasks effectively. As a result, the working stage can be an exciting, challenging, and deeply satisfying time in the internship.

A possible pitfall of the working stage is that students can lapse into carrying out routine duties, enjoying their greater comfort and confidence, without exploiting the learning opportunities that they might pursue. If the learning agreement has been too modest, you might even find that all or most of your learning goals have been accomplished at this early juncture. If this situation develops in your internship, you and your supervisor should revise the learning agreement to ensure that your time and activities are wisely invested.

A second potential pitfall of the working stage may be the tendency for students to begin working too independently. As an intern becomes more capable, the supervisor might tend to spend less time with the intern, resulting in less supervision, less teaching and, as a result, less learning. An intern in such a situation might be quite flattered, even relieved, to be more frequently away from the supervisor's watchful eye. Nevertheless, for the sake of your own learning, if this situation should develop you should politely express your concern about the decreased supervision time to your supervisor.

A Student's Reflections on the Internship's Development

After four weeks on the job the new has worn off. Let's face it—it is no longer fun to get up at 6:00 A.M. every day. There's nothing wrong with my internship, but it has just become more routine now. My roommates are still on college student schedules, staying up till all hours. I'm falling into bed exhausted by 10:00 P.M. I must admit, though, that there are some rewards for hanging in there. I'm much more comfortable at my placement than I was at first, and I feel good that they trust me to do more on my own. My supervisor also told me that a lady who called and asked me some questions over the phone had commented to her about how polite and professional I was. It's nice to know that some of my strengths are beginning to show.

Stage 4: Termination Stage

The internship shifts into its final stage when the student and supervisor begin to anticipate and plan for the end of the internship. New assignments are not as freely given due to time constraints, and the work that the intern is currently doing must either be completed or plans made to transfer responsibility to other workers. As you bring the work of your internship to a close, you must say good-bye to supervisors, colleagues, and clients. Depending upon the nature of these relationships, this can be a time of great emotional intensity. At the same time, however, the termination stage can also be a period of equally great satisfaction and even pride.

The final stage of the internship is also a time for reflection. You might reflect upon such questions as: How well did I perform in this internship? What have I really accomplished here? What have I learned after all? What do I have regrets about? What do I value and appreciate about this internship? Where will I go from here in my professional development? Such reflection is a positive and valuable aspect of your experience and will help you to engage more meaningfully in the formal evaluation process with your field supervisor that occurs toward the end of this stage.

The quality of your experience during the termination stage will depend in part upon your own patterns and skills in dealing with endings. The termination stage will give you the ideal opportunity, of course, to examine and work on these patterns and skills. Therefore, many students find it useful for supervision during this stage to focus specifically on the skills involved in productive endings. Even though you might have become very competent and independent in handling various responsibilities during your internship, you might find that you need closer guidance and teaching in order to deal effectively with the new challenges of ending relationships.

As implied above, pitfalls during this stage tend to relate to the student's patterns in handling endings. Therefore, it is wise for students to evaluate these patterns prior to entering the termination stage. Some students have great difficulty letting go. These students might delay transferring tasks or clients to other workers, hanging on to these responsibilities as long as possible. Other students might tend to minimize the significance of the ending, repeatedly telling everyone, "I'll be back to visit." Denying that the end is coming can block the student from saying good-bye in a real and meaningful way and can even give clients the false reassurance that the student will still be available. Before making such casual promises, students must examine closely the appropriateness of such visits, their motives for wanting to stay in touch, and the agency's policies about continued relationships with clients. A third pattern that sometimes emerges as students approach the end of the internship might be described as "premature disengagement." As the student begins to anticipate leaving the organization, attention to the tasks at hand may decrease. Within this pattern, it is as though the student has emotionally and mentally left the internship before completing it. A student who has earlier been hard-working, motivated, and attentive to detail can lapse into daydreaming and distractibility as the end of the internship approaches. Obviously, this is a behavior pattern that can only harm the quality of a student's performance and the supervisor's evaluation of the student. Although it is challenging to work on letting go and staying engaged at the same time, this is exactly what the termination stage of the internship calls for. Throughout the termination stage, you should stay meaningfully engaged and dedicated to the people and tasks that are yours to respond to each day.

Positive termination is an important and challenging task that will need careful consideration and planning. Chapter 12, "Leaving Your Internship," will help you meet this challenge more effectively by dealing with termination issues in greater detail.

A Student's Reflection on Termination

Today was my last day. It was a good day but also depressing. This internship has been a lot to take in, but I learned so much. The people I worked with always took time to talk with me after group and other events to see what I thought and how I felt about what had happened. I appreciate the time they took with me and that they asked for my ideas instead of just telling me their's. The internship also made me think a lot about my own issues and my family, and, even though I never discussed it with anyone, I feel that I was able to put into perspective some personal issues. This was an unexpected but wonderful benefit. Now for the depressing part. I have learned and grown so much that I hate to leave. I will really miss the people and feel just plain sad about the fact that it's over. I am thankful that I will have opportunities to do more fieldwork in this major in the future. That makes it easier to move on.

Exercise 1.4

Discuss your internship in terms of its development thus far. What particular developmental tasks have you already dealt with in your internship? Which ones are you currently working on? How do you see your experience thus far as fitting into the developmental stages of the internship described above? What particular pitfalls discussed in this section are you able to identify with as relevant to your experience thus far?

CONCLUSION

Clarifying expectations is an important step in getting your internship off to a good start. You, your field supervisor, and your faculty liaison have certain expectations of the internship. Knowing that your internship will develop through some predictable stages, as well as being familiar with the issues and tasks inherent in each stage, will better prepare you to meet the demands of the internship over time. Now that you have a better understanding of what to expect in your internship, you are ready to move to the next step, Chapter 2, "Getting Started."

Your Reflections on Getting Ready for Your Internship

REFERENCES

Cantor, J. (1995). *Experiential learning in higher education: Linking classroom and community.* Washington, DC: ERIC Clearinghouse on Higher Education.

Chiaferi, R., & Griffin, M. (1997). *Developing fieldwork skills.* Pacific Grove, CA: Brooks/Cole.

Council for Standards in Human Service Education. (1995). *National standards for human service worker education and training programs.* Fitchburg, MA: Author.

Dore, M., Epstein, B., & Herrerias, C. (1992). Evaluating students' micro practice field performance: Do universal learning objectives exist? *Journal of Social Work Education, 28,* 353–362.

Fry, R. (Ed.). (1988). *Internships: Advertising, marketing, public relations & sales.* Hawthorn, NJ: Career Press.

Gladding, S. (1992). *Counseling: A comprehensive profession* (2nd. ed.). New York: Macmillan.

Grant, R., & McCarty, B. (1990). Emotional stages in the music therapy internship. *Journal of Music Therapy, 27*(3), 102–118.

Grossman, B., Levine-Jordan, N., & Shearer, P. (1991). Working with students' emotional reaction in the field: An educational framework. *The Clinical Supervisor, 8,* 23–39.

Inkster, R., & Ross, R. (1998). Monitoring and supervising the internship. *NSEE Quarterly, 23*(4), 10–11, 23–26.

Kerson, T. (1994). Field instruction in social work settings: A framework for teaching. In T. Kerson (Ed.), *Field instruction in social work settings* (pp. 1–32). New York: Haworth Press.

Lamb, D., Barker, J., Jennings, M., & Yarris, E. (1982). Passages of an internship in professional psychology. *Professional Psychology, 13,* 661–669.

Michelsen, R. (1994). Social work practice with the elderly: A multifaceted placement experience. In T. Kerson (Ed.), *Field instruction in social work settings* (pp. 191–198). New York: Haworth Press.

Pettes, D. (1979). *Staff and student supervision.* Boston: George Allen & Unwin.

Siporin, M. (1982). The process of field instruction. In B. Sheafor & L. Jenkins (Eds.), *Quality field instruction in social work* (pp. 175–198). New York: Longman.

Sweitzer, H., & King, M. (1999). *The successful internship.* Pacific Grove, CA: Brooks/Cole.

Young, M. (1998). *Learning the art of helping: Building blocks and techniques.* Englewood Cliffs, NJ: Prentice-Hall.

Getting Started

As discussed in the previous chapter, a tremendous amount of work must be done to get your internship underway. Even before your first day on the job, there is much for you to do. Simply selecting a fieldwork site can be time-consuming and complicated. If your program gives you input on your fieldwork site, there are many variables for you to consider in making this choice. The discussion and exercises in this chapter will engage you in this decision-making by asking you to think about many of these variables carefully. The chapter will also give you the information you need to plan for and participate in the initial meeting with your supervisor. Finally, once your fieldwork has begun, one of your most important tasks will be developing a learning plan. This chapter will acquaint you with the fieldwork learning plan and will take you through the step-by-step process of developing one.

A Student's Reflections on Getting Started

It's hard to order when you don't know what's on the menu. That's how I feel when I try to decide where to do my internship. The human service field is so broad and varied, it is impossible to know about all the opportunities and careers that are out there. Where do I start? Sometimes I think that practically everything in this field interests me so what does it really matter where I end up. Other times I get nervous about it and feel like this internship is setting the stage for the rest of my professional life. Whatever field I get experience with now will help me get a job in that field later. So I go back and forth between thinking that where I do my internship is the most important decision in the world and thinking that almost any place would be a good start.

SELECTING A FIELD SITE

Finding a field site that fits your learning needs is what selecting a field site is all about. Human service programs use a variety of methods for making student placements for fieldwork. In some cases, the student is entirely responsible for securing the placement and the faculty member makes a contact with the agency to review the placement's appropriateness. In other cases, faculty members take responsibility for placing students, taking into consideration student interests, goals, and skills. In most cases, students have at least some input into their placement decision (Simon, 1999), but in some programs students may be assigned to placements with little to no input (Royse, Dhooper, & Rompf, 1996). Students can learn a great deal even before the internship begins by being actively involved in the selection of a field site. Developing greater awareness of your career interests, building your job search skills, and enhancing your ability to communicate clearly and assertively are just a few obvious examples of learning that can take place in the field site selection process (Bertea, 1997).

Those programs that allow student input often ask students to submit information in writing as to the type of placement they would like to secure. It is not uncommon for student requests to be very general, for example, "I want to work with children" or unrealistically specific, for example, "I want to work in a residential treatment program with teenage substance abusers, and it needs to be within walking distance of the campus." These examples illustrate some of the most obvious considerations when thinking about a field site, such as the population served and the field of practice. You will be able to make a more informed decision about your field site by considering these as well as other factors more thoroughly.

A Student's Reflections on Choosing a Field Site

I have always wanted to work with children who have handicaps. I have said this for so many years, I think it might have just become habit. Now that I have a chance to test out some different careers, I am drawn to trying something different. I have worked at a camp for children with handicaps for the past few summers so I have some idea of what that type of work is like. While I'm a student and have this opportunity to do Practicum and Internship, I would like to broaden my horizons. I would be interested in trying out an administrative role of some kind. Also I think that working with the elderly could be very gratifying. The mental health field interests me a lot, too. I was in counseling at one point, and it helped me so much personally. I wish I could do a dozen different internships.

Considering Populations

Human services agencies work with people in need, providing services throughout the life span from birth to death. Due to specialization within the field, most agencies have a specific segment of the population as their target group. Many students identify the work that they wish to do in the human service field by identifying the age group with which they are most comfortable. This selection is often based on the student's previous job or volunteer experience, experiences with family members, or experiences in the student's own personal life. Although all of these form a good basis from which to consider various options, it is helpful to think more carefully about these options rather than choose a given population based on a general sense of comfort. As students are exposed to various populations, it is not uncommon for them to find that they enjoy working with a group that they had not considered before or perhaps had even avoided. Such experiences suggest that students would be wise to approach the process of field site selection with as open a mind as possible. The following exercise will help you to think through thoroughly and systematically the possibility of working with various populations. If you already have a field site, the exercise will help you to envision various scenarios that you might pursue in future internships, volunteer work, or employment.

Exercise 2.1

Consider each age group identified below and identify (1) the potential rewards of working with this group and (2) the potential frustrations or difficulties in working with this group. Briefly write down your thoughts and reactions as you respond to these questions. Although some groups will be initially more appealing to you than others, try to develop a balanced reaction to each population by responding to both (1) and (2) above.

Preschool Children (0–5 years)

Grade School Children (5–12 years)

Adolescents (13–18 years)

College-Age Young Adults (18–22 years)

Adults (23–65 years)

Older Adults (65–85 years)

Oldest Adults (>85 years)

Considering Fields of Practice (Settings)

Human services agencies vary broadly in their missions. An important step in selecting your field site is identifying the human problems or needs that you would like to learn more about. This dimension of an agency's or a professional's identity is referred to as the *field of practice* or *setting* (Barker, 1996). The field of practice shapes a worker's experience considerably as specialized expertise is required in each field. Given the range of human experience, it is impossible to identify all of the fields of practice here. The following exercise will list several fields of practice for you to consider and react to as you develop your ideas about a potential field site.

Exercise 2.2

Think about your reaction to each field of practice listed below and briefly record your thoughts. Try not to focus on whether you already have the knowledge and skills to work in this field. Remember that you are entering the setting to gain these. As you consider each field of practice, respond to the following: (1) What interests you or attracts you about this field of practice? (2) What concerns, fears, or reservations do you have about working within this field? (3) Record your overall motivation to learn more about this population, using the ratings of "Very High," "High," "Moderate," "Low," "Very Low."

Poverty

Homelessness

Child Abuse and Neglect

Substance Abuse/Addictions

Mental Illness

Developmental Disabilities/Mental Retardation

Physical Disabilities

Education

Health Care

Crisis Services

Domestic Violence

Concerns Related to Aging (e.g., social isolation, physical/mental decline, retirement)

Crime

It would be useful to discuss your responses to Exercise 2.1 and Exercise 2.2 with other students and to have your work reviewed by a person who is experienced in the human services field, perhaps one of your faculty members. Sometimes initial reactions to various fields are based on misinformation or stereotypes. Sharing your thoughts with others will help to identify misconceptions, biases, or false assumptions that might be present.

As you select a field site, you will probably want to examine most closely those fields above that you responded to with at least "moderate" motivation. For each of these fields, think at greater length about what aspects of that field capture your interest. In some cases, human service workers are drawn toward a particular field in an effort to deal with unresolved issues in their own personal lives. For your own well-being, as well as that of your clients, you should not select a field site that will take you too deeply into troublesome issues that may be present in your own life. In such a situation it is often difficult to sort out your own thoughts and feelings from those of your clients and to maintain appropriate boundaries and objectivity in your helping relationships. If this is a field that you want to enter, you may still do so after you have resolved your own issues either through professional counseling or personal reflection and effort. If you are uncertain as to whether you have resolved an issue sufficiently to begin professional work in that area, discuss your concerns with a trusted faculty member or a professional counselor.

Considering Agency/Worker Function

Human services agencies perform a variety of roles and functions in relation to any given population and field of practice. Agency function has to do with the nature of the goals and interventions within the organization. The local community often has a number of agencies working in different ways to benefit a given population. Also, within larger human services agencies there are typically specialized departments performing a variety of tasks to benefit the target population. For example, within a given community there may be a mental health center that provides direct services to clients through its counseling services, while another agency, perhaps the mental health association, provides community education and prevention programs. There may be a department within one of these organizations that leads fund-raising efforts and carries out other administrative and leadership functions to expand services for the mentally ill within the community.

In 1969, the Southern Regional Education Board conducted a study of human services workers that identified 13 different functional roles of human services workers (Mandell & Schram, 1985; Mehr, 1998; Neukrug, 1994; Woodside & McClam, 1998). Some agencies will conduct many of these functions, while others may conduct only one or two of them. Considering each functional role below will help you to identify more clearly the type of experiences and skills you are seeking through your fieldwork.

Exercise 2.3

As you consider each of the following functional roles, write a response in which you identify: (1) your level of interest in this role, (2) your current level of skill in performing this role, and (3) how important you think it is to develop further skills in this role prior to completing your human service program.

Outreach—going into the community to work with clients in their environment, especially for the purpose of reaching under-served populations

Broker—helping clients locate and access services

Advocate—assertively seeking services for clients and defending clients' rights

Mobilizer—organizing community groups to develop needed services and/or initiate social change

Behavior change—using intervention strategies and counseling skills to facilitate client change

Teacher/educator—informing individuals and groups

Caregiver—offering direct care to clients physically, emotionally, and/or socially

Evaluator—assessing human services programs for effectiveness and financial accountability

Assistant to specialist—working closely with a highly trained professional as an aid in delivering services

Data manager—gathering information and organizing it for use in a variety of tasks involving direct services to clients and/or administrative services

Administrator—supervising human services agencies or programs

Community planner—identifying community needs and designing programs to meet those needs

Consultant—offering knowledge and expertise to other professionals and/or to client and community groups

Reflecting upon these functional roles will help you to identify the specific roles you would like to enhance in your own professional development. Although students usually begin thinking about their fieldwork by identifying the population or field of service they would like, it may be more useful in some cases to think first in terms of functional roles. These roles are generalizable from one population and field of practice to another. For example, advocacy skills learned through work with welfare clients can be easily transferred to working with abused children, the mentally ill, or any other client group. Such skills form the core of your development as a professional and, therefore, may well be of greater significance and lasting value than gaining experience with a given population or field of practice.

Exercise 2.4

Review the writing you have done in Exercises 2.1, 2.2, and 2.3. Keeping in mind your reactions to the various populations, fields of practice, and functional roles, describe at least three (3) field placements that would be beneficial for you at the present time. Avoid simply naming agencies or organizations that you might know about. Instead, describe the type of learning experiences that you need. Draw upon the concepts of population, field of practice, and functional roles in your descriptions.

Faculty members and more advanced students in your program can help you identify organizations that offer the types of experiences which you describe. Also, depending upon your school's policies regarding making placements, you might contact a few agencies regarding your interests. The work you have done in the exercises of this chapter will help you articulate your learning needs and give you some ideas about where to start in your search for a field site.

THE INITIAL MEETING WITH YOUR PROSPECTIVE SUPERVISOR

Once you and the faculty of your program have secured a tentative site for your fieldwork, you should schedule a meeting with your potential field supervisor (Faiver, Eisengart, & Colonna, 1995; Morrisette, 1998; Thomlison et al., 1996). This meeting is in your best interest because it helps to ensure that there is a good match between what the agency has to offer and your interests. Also, most supervisors want to interview students prior to approving the placement officially. It is generally the student's responsibility to initiate and schedule this meeting.

Typically, the supervisor will want to focus on such issues as why you are interested in this particular site, what your career goals are, what experiences (both classroom and direct practice) you have had that might be relevant to the internship, any previous field placements you might have had, and what you are hoping to gain from the field experience. Additionally, you should give thought to questions that you would like to ask during this meeting. Students are generally interested in the nature of the client population served within the agency, the range of services and programs offered, the types of experiences the agency will make available, and typical expectations of students. Thomlison et al. suggest a number of questions students might ask their field supervisors during the initial meeting, including the following:

> What are the kinds of things I will be doing and learning here? . . . What systems will I be working with? . . . Will I have more than one [field] instructor? . . . What is the largest single problem or issue facing your organization (e.g., agency or setting) that I should know about? (1996, p. 47)

In addition to such questions, you might also want to raise a number of general orientation questions, such as:

- What is a typical day like in this setting?
- How should I dress for the internship?
- What hours will I be working?
- Where will I be located physically in the facility?
- What arrangements, if any, do I need to make for lunch or other meal times?

No doubt you will have additional questions based upon your own particular interests and concerns.

The initial interview is similar to a job interview in many ways, but it is usually more spontaneous and informal (Wilson, 1980). The meeting is somewhat like a job interview, however, because both you and the supervisor are deciding if a placement in that particular setting is right for you. As in a job interview, you will want to make a good first impression by arriving on time, dressing appropriately, making eye contact, and shaking hands firmly. In the interview you will, of course, want to convey your eagerness to learn and your interest in the organization while also conveying something of your own abilities and relevant background. If you have had previous experience in human services, it would be a good idea to prepare a resume reflecting these experiences and bring a copy for the supervisor.

Based upon this interview the placement generally moves toward formal approval. If either you or the supervisor has serious reservations about the placement following this discussion, your faculty liaison should be notified as soon as possible. The concerns raised by the interview may be resolvable, but if not, an alternative placement will need to be arranged. To allow sufficient time for this possibility, it is important to schedule your initial meeting with the field supervisor in a timely manner.

A Student's Reflections on Meeting His Supervisor

I thought I was going to meet my supervisor just so we could start getting to know each other. Our conversation was more like a job interview than I expected it to be. For example, she asked about my educational background and practical experience, and she asked what I see as my strengths and weaknesses. Although it made me a little nervous, I see it as good practice for my job interviews later on. I am excited about the possibilities we discussed for my in-

ternship. It sounds like my supervisor is going to trust me and give me some real responsibility. After a few weeks I will even be able to have my own cases to work with independently. I do, however, have some concerns. The agency, and my supervisor in particular, seem to be very busy. I'm concerned that they might not have a lot of time for a student. I have given it some thought and have decided that I do want the placement nevertheless. I have no doubt that I can learn a great deal there, even if people are very busy. I learned so much just by going for a visit. Another smaller concern that I have is about how to dress. My supervisor was really dressed up, and I noticed that several other workers were as well. Most of the male staff members were wearing ties, and the women were wearing dresses and heels. Am I expected to dress like that? I meant to ask at the time, but it slipped my mind with all the other things to discuss. I plan to call her before my first day to ask about this.

Exercise 2.5

In preparation for the initial meeting, write in the space below the questions you would like to ask or topics you wish to discuss in this meeting. Also include any information about yourself and your experiences that you would like to share with your supervisor.

An initial meeting with your supervisor in which your questions are answered will help you to arrive on the first day of the internship with greater confidence and comfort. It is well worth the time and effort involved as it helps you make a smooth transition into your new environment. Following this initial meeting, it is a good idea to follow-up with a thank-you note to your field supervisor (Faiver et al., 1995). The note can either indicate your interest and eagerness to begin the internship or can convey that you have decided to pursue a different placement more consistent with your goals. In either case, your conversation with this human services professional was helpful to you in making a decision about your future and is worthy of a word of thanks.

DEVELOPING A LEARNING PLAN

The initial meeting with your supervisor begins to lay the groundwork for your field-work learning plan. The learning plan is a written document in which your learning goals are specifically identified as well as the strategies or activities for accomplishing each goal (Bogo & Vayda, 1995; Thomlison, et al., 1996; Wilson, 1980). The learning plan also identifies how your progress toward each goal will be measured. This plan may be referred to within some human service programs as a *learning contract* or *learning agreement*. Whatever term is used, this document is an essential tool for guiding your fieldwork. In field experiences that are a full-academic year in duration, the program may allow as much as a month to develop the plan. In experiences that are a semester or less, the plan must be developed more quickly, usually within one to two weeks. Once developed, the learning plan serves as an important road map, directing and organizing your experiences. Therefore, it is important that you develop your learning plan as early as possible. In most programs, the learning plan is open to revision throughout the field experience. Contingent upon the mutual agreement of the student, field supervisor, and supervising faculty member, goals and strategies can be modified, added, and/or deleted within reason. As the placement progresses, new opportunities often arise and new interests develop. By the same token, there are also times in which previously anticipated opportunities do not develop. While the integrity of the learning plan needs to be preserved in order for it to serve its purpose, it must also be a living document, capable of responding to changing circumstances.

A Student's Reflections on the Learning Plan

At first I looked at writing up a learning plan as busy work. Now that it's done I definitely see the value of it. It helps me keep in mind what I'm supposed to be getting out of this. I think without the learning plan I wouldn't be nearly as clear about why I'm here, what I'm supposed to be learning, and what I'm supposed to be doing. I think it has also helped the staff see me as a student and not a volunteer. They take me more seriously, I think, because it's clear that I'm here to learn.

Identifying Goals

The development of the learning plan is a mutual process involving you and your field supervisor. Forming the basis of the plan is the discussion in your initial meeting with your supervisor in which you explained something about your interests and the field supervisor described the experiences that the agency could offer. You and your field supervisor now need to discuss in more specific terms the exact nature of your learning goals. Although it is your responsibility to identify your own learning needs and to generate your own ideas about your specific goals, your field supervisor's involvement is needed to ensure that these goals are realistic for the particular setting and for the time available. Additionally, as an expert in the field, the field supervisor can ensure that you do not omit important areas of learning.

The level of student responsibility for shaping the fieldwork learning experience often requires a major adjustment for students. In a traditional course, the professor identifies course objectives, makes assignments, and informs students as to how they will be evaluated. Developing the learning plan requires that students take far more leadership in shaping their learning process. In creating the learning plan, you will initially produce a draft of your goals, the strategies that you will use to accomplish them, and the methods that will be used to measure your progress. Your field supervisor can then review this draft and make suggestions to further shape and develop the plan. You should take the time to develop your learning plan carefully, not only because it will direct your learning in the internship but also because it can be used for legal purposes in some circumstances (Gross, 1993).

To develop your learning plan, you might begin by reading the general goals for the internship that have been developed by your human services program. These goals are generally written in either the syllabus for the internship or in a student handbook or both. The program's goals for the internship will help you to identify the broad parameters within which your learning should operate. Some programs direct students to set goals in specified areas. Structuring the goal-setting process in this way is helpful because it guides students to develop a well-balanced plan for their learning. For the purposes of the exercises in this chapter, you will develop goals related to the following three categories of learning:

1. **Knowledge** defined as learning and understanding facts, concepts, theories, information, or ideas.
2. **Skills** defined as the ability, gained through practice, to do things well.
3. **Personal development** defined as attitudes, values, biases, and/or habits that are uniquely your own.

Goals in each of the preceding three areas should be as specific as possible. Keep in mind that a goal is an outcome or an endpoint, not an activity, program, or event. To state each goal as an outcome, ask yourself exactly what knowledge, skills, or personal development you hope to have attained by the end of the internship. For example, "I will become familiar with the statutes in North Carolina pertaining to domestic violence" is a good goal statement. "I will attend a conference on domestic violence" is *not* a goal statement but a statement about an intended activity.

Exercise 2.6

Step 1: Identify Goals

A well-developed learning plan will identify several goals for each category of learning (knowledge, skills, and personal development). For the purposes of this exercise, identify at least two goals for each of the three categories of learning listed and defined above. Feel free to develop additional goals.

Knowledge:

Goal 1:

Goal 2:

Skills:

Goal 1:

Goal 2:

Personal Development:

Goal 1:

Goal 2:

Identifying Strategies

Having set your goals, you are now ready to consider the strategies that you will use to accomplish each goal. When identifying strategies, think of specific activities that will enable you to achieve each goal that you have listed. As with your goals, your strategies will be most useful if they are as specific as possible. Strategies might include such activities as observing, attending meetings, interviewing co-workers, reading, and so on. Continuing to develop the example discussed previously, a goal with strategies might look like this:

Goal 1: I will become familiar with the statutes in North Carolina pertaining to domestic violence.

Strategies:

a. I will attend a state conference on legal issues affecting women to be held in Raleigh, NC on March 24–26.

b. I will observe staff members as they inform domestic violence survivors of their legal rights.

c. I will attend court and observe its proceedings each Friday morning.

d. I will read *Your Legal Rights: A Guide for Survivors of Domestic Violence*, by Olivia Kramer.

Exercise 2.7

Step 2: Identify Strategies

Using the goals you developed in Exercise 2.6, identify strategies that you can use to achieve each goal. Each goal will probably have several strategies.

Knowledge

Goal 1:

Strategies:

Goal 2:

Strategies:

Skills
Goal 1:

Strategies:

Goal 2:

Strategies:

Personal Development
Goal 1:

Strategies:

Goal 2:

Strategies:

Identifying Methods of Measurement

As the last step in developing your learning plan, you and your field instructor will specify ways to measure your progress. Ideally, the ways you select to measure your progress will rely upon information that is directly observable by your field supervisor and/or your faculty liaison since they share responsibility for evaluating your work. The methods you use may vary with each goal. For some goals you might use only one method, while for other goals you might use several.

Continuing to develop the earlier example, the methods of measurement for the goal, "I will become familiar with the North Carolina statutes pertaining to domestic violence," might include such methods as, "My supervisor will observe as I explain to clients their legal rights and explain how the legal process works," and/or "I will write a newsletter article regarding the legal rights of domestic violence victims that my supervisor will read and evaluate," and/or "I will make a presentation on legal issues pertaining to domestic violence in my field seminar class that my faculty liaison will evaluate." All three of these examples involve methods by which the field supervisor, faculty liaison, or both can directly observe and give feedback to the student.

Measurement methods often include such activities as direct observation of student work, supervisor review of written documents that the student has produced, supervisor review of student videotapes or audiotapes, and/or supervisor or faculty liaison observation of student presentations. Also, students often keep journals in which they document what they are learning. This, too, might be reviewed and evaluated by the supervisor and/or faculty liaison as a method of measurement for particular goals. As you construct your learning plan, your field supervisor or faculty liaison might have additional suggestions regarding how to evaluate your progress on the various goals that you set.

Exercise 2.8

Step 3: Identify Methods of Measurement

To complete your learning plan, identify methods of measurement for each goal you have set. Try to identify 2 to 3 possible methods of measurement for each goal. Although you may not include all of these methods in the final draft of your plan, they will provide a good starting point for discussions with your field supervisor.

Knowledge

Goal 1:

Method(s) of Measurement:

Goal 2:

Method(s) of Measurement:

Skills
Goal 1:

Method(s) of Measurement:

Goal 2:

Method(s) of Measurement:

Personal Development
Goal 1:

Method(s) of Measurement:

Goal 2:

Method(s) of Measurement:

Through your work on Exercises 2.6 to 2.8, you have created a beginning draft of your learning plan. You will probably want to set additional goals for some areas. As you do so, continue to use the process and steps that you have learned in this chapter. When you have completed a draft of your learning plan that represents your best and most complete effort, make a copy of it for your field instructor and schedule a time to discuss it. This conversation will probably produce modifications and revisions, leading to your final draft. Both you and your field supervisor should sign the learning plan in its final form to indicate your mutual support for the plan and your intention to implement it.

Once your learning plan is complete, you will find that you feel more focused in your work and that you can better articulate what you are learning from your fieldwork. As you implement your learning plan, discuss with your supervisor how well the plan is serving your learning needs. If you feel that the plan needs modification as you go along, express this to your supervisor as well. The learning plan should always serve to enhance your learning by providing focus, direction, and structure.

CONCLUSION

Beginnings are important. This chapter has provided much of the information you need to launch your internship successfully. Considering carefully your learning needs and interests as you select a potential field site, exploring important issues in your initial meeting with your field supervisor, and developing a well-thought-out learning plan are essential steps in laying a solid foundation for your internship.

Your Reflections on Getting Started

REFERENCES

Barker, R. (1996). *The social work dictionary.* Washington, DC: NASW Press.

Bertera, E. (1997). Empowering human services students to select their practica: Insuring goodness of fit. *Human Services Education, 17*(1), 65–67.

Bogo, M., & Vayda, E. (1995). *The practice of field instruction in social work: Theory and process.* Buffalo: University of Toronto Press.

Faiver, C., Eisengart, S., & Colonna, R. (1995). *The counselor intern's handbook.* Pacific Grove, CA: Brooks/Cole.

Gross, L. (1993). *The internship experience* (2nd ed.). Prospect Heights, IL: Waveland.

Mandell, B., & Schram, B. (1985). *Human services: Introduction and interventions.* New York: Macmillan.

Mehr, J. (1998). *Human services: Concepts and interventions.* Boston: Allyn & Bacon.

Morrisette, P. (1998). The undergraduate preinternship process: Issues and recommendations. *Human Service Education 18*(1), 49–55.

Neukrug, E. (1994). *Theory, practice, and trends in human services: An overview of an emerging profession.* Pacific Grove, CA: Brooks/Cole.

Royse, D., Dhooper, S., & Rompf, E. (1996). *Field instruction: A guide for social work students* (2nd ed.). White Plains, NY: Longman.

Simon, E. (1999). Field practicum: Standards, criteria, supervision, and evaluation. In H. Harris & D. Maloney (Eds.), *Human services: Contemporary issues and trends* (pp. 79–96). Boston: Allyn & Bacon.

Thomlison, B., Rogers, G., Collins, D., & Grinnell, R. (1996). *The social work practicum: An access guide.* Itasca, IL: Peacock Publishers.

Wilson, S. (1980). *Field instruction.* New York: Free Press.

Woodside, M., & McClam, T. (1998). *An introduction to human services* (3rd ed.). Pacific Grove, CA: Brooks/Cole.

Getting Acquainted

As a newcomer you have an ideal opportunity to gather information about your field-work organization. You will be introduced to many staff members, most of them eager to tell you about the agency and to answer your questions. You will probably be given a general orientation to the agency and offered printed materials to peruse. Also, in the first few days of your internship, you will likely have some unstructured, unscheduled time that you can use to get acquainted with the organization and the people within it. These common rituals and rhythms of getting started open the door for your exploring and learning as much as you can about your field site. In some circumstances, the nature of the work may be such that you do not have the luxury of a slow startup or a thorough orientation. In this case, you will need to rely on your own observations, assertiveness, and resourcefulness to get oriented. In any case, this chapter will help you to focus and direct your attention toward the specific aspects of agency identity that make each organization unique and interesting.

A Student's Reflections on Getting Acquainted

My entry into the internship has been one of mixed emotions. At times, I've felt alone, as though I am a stranger looking in. The counselors don't really have time to play "nice to meet you." The residents come first, and I realize that this is how it should be. Nevertheless, I have jumped in with both feet and am learning a lot about the place as I go along. My endless questions are always answered, and no one is afraid to give me directions or instructions. Overall, I have been welcomed with open arms. But like any new job, it takes a while to learn the ropes. I have to realize that I can't learn everything in one day. I'm glad to be here. Every morning I am faced with a challenge that I look forward to conquering.

GETTING TO KNOW YOUR AGENCY

Getting to know as much as you can about your agency early on will not only help you feel more comfortable and confident in your new setting, it will also help you to understand and integrate your future experiences as they unfold during your internship.

Following is a list of some of the most useful and accessible means of gathering information about your field site in the earliest days of your placement. Discuss with your field supervisor your interest in learning more about the agency and determine which of the following sources of information you will have access to in your organization:

- Read pamphlets, brochures, or any other written information for distribution to the community.
- Read the agency policy manual.
- Read any periodic reports that the agency prepares for its governing and/or funding bodies—for example, budget proposals, annual reports, grant proposals, and so on.
- Formally interview or informally speak with key people in the organization, such as the agency director, the human resources director, department heads, and employees with especially long seniority.
- Attend any of the routine meetings of the organization that you are allowed to attend.

In addition to these ideas, your field supervisor might have additional resources to suggest. Of course, one of the best ways to learn about your organization is by simply being a good observer. Keeping your eyes and ears open and your mind engaged at all times is your best information gathering tool.

Exercise 3.1

Describe your first impressions of your field site. Describe its physical appearance both inside and out as you made your first visit. What messages and values seem to be conveyed to clients through these physical characteristics? How were you met and greeted by the staff in the organization? What were your observations about how clients are met and greeted? What messages and values are conveyed through these behaviors?

GETTING TO KNOW YOUR CO-WORKERS

In your first few days in the organization you may be overwhelmed with new faces and new names, especially if your organization is large. Allow yourself some time to simply observe the various people working around you and their interactions. It is, of course, understandable that you might be nervous at first about approaching your co-workers for conversation. Try to identify those workers who may be especially approachable and take a few risks in extending a hand and introducing yourself. You will, of course, want to approach those workers whose job, education, or professional background particularly interest you in some way. Gathering the information for the exercises in this chapter will give you a good focus and an incentive for approaching your co-workers. Most professionals are pleased, even flattered, to find that a student is interested in their work. Nevertheless, in approaching staff members, be respectful of the fact that they may not have time for a lengthy conversation at that particular time. It may be best to ask for a scheduled time to meet if you find that you have several questions to discuss.

A Student's Reflections on Getting Acquainted

Every day I learn more and more about this place. There is so much more here than meets the eye at first. I never knew, for example, that this organization was started by the Lutheran Church (my own denomination). Also I recently met an employee who is running for public office because she wants to do more to influence social policy and human service funding in the state. She has opened my eyes to how government policies on different things affect the quality of life. I have never been interested before in government policies on issues like minimum wage, child care, and health insurance. This internship is really helping me to see and understand the "big picture."

As a general rule, you will enhance your experience in your internship if you try to stay in the flow of interaction within the organization, getting to know as many people as possible. If the staff has a gathering place for lunch and breaks, try to be there for those times as much as possible. If you have your own office, keep the door open rather than closed whenever you can. If your supervisor tells you that you may attend a certain meeting "if you would like," attend it if at all possible, even if it is at an inconvenient hour. Make yourself as approachable and involved as possible, remembering that staff members, too, may have some anxieties and uncertainties about approaching you.

As you make connections with your co-workers, you are engaging in the important, but generally untaught, career skill called *networking*. Networking is the art of using your interpersonal skills to forge links with others. Networking may be defined as "the relationships professionals cultivate with other professionals to expedite action through the social system" (Barker, 1996, p. 253). In human service organizations, like all organizations, a great deal occurs through informal channels. Having good relationships with colleagues helps you to provide better services and linkages for clients as they move through the human services system. These relationships can also provide excellent sources of informal mentoring, support, and guidance for your own professional development and future job search (Neukrug, 1994).

In getting to know your co-workers, do not overlook support staff. Secretaries and administrative assistants often have an excellent overview of the organization and know a great deal about interacting with clients based on years of experience. Maintaining a positive relationship with support staff will also enable you to do your job better due to the support and information that they can provide (Thomlison et al., 1996).

Exercise 3.2

Identify at least one staff member whom you have met and talked with in the first days of your fieldwork. Identify the individual by name and job title, summarize your conversation, your impressions of the individual, and your reaction to the conversation. Also identify particular people in the organization whom you have not met but plan to meet within the next few days or weeks. In identifying these individuals, think of gaps in your current knowledge and understanding of the agency. What perspectives are you missing?

LEARNING ABOUT THE AGENCY'S MISSION, OBJECTIVES, GOALS, AND STRATEGIES

A clear statement of an organization's mission, objectives, goals, and strategies is central to its efficient functioning. Your understanding of your field site will be enhanced by examining these elements of its identity. These elements of the agency's identity may be predetermined through legislative or funding mandates or may be determined more locally by the staff, administrators, and/or board of directors. Human service agencies often have a formal, written mission statement in the first pages of their policy manual. The prominence of this statement reflects its importance to the agency's identity. The mission statement of an agency sets out the rationale for its existence, explaining whom the agency is intending to serve and the community needs or problems that it is seeks to address. The mission statement also conveys important information, directly or indirectly, about the agency's philosophy of helping and the values that underpin its efforts.

Closely related to the agency's mission are its objectives and goals as well as the strategies it uses to meet them. The agency's objectives can often be found in close proximity to its mission statement. The objectives of the agency describe the outcomes that it is trying to accomplish, often using broad, difficult-to-measure terminology. The objectives are effective in giving the agency direction toward the long-term achievement of its purpose. In contrast, agency goals are written or discussed in more specific, measurable language and are effective in focusing and directing more immediate efforts. For example, an agency may have the objective of assisting the homeless in the community and the goal of reducing the number of homeless people in the community by 10% by the year 2005 (Barker, 1996). Many agencies have written goals and objectives for the organization as a whole as well as for each department of the agency, while other agencies are less formal and comprehensive in their administrative methods.

Agencies employ specific strategies to fulfill their missions and accomplish their goals and objectives. Programs, services, and activities are planned, designed, and implemented toward this end (Weiner, 1990). The strategies include not only the services and programs directed toward clients but also such "behind-the-scenes" administrative efforts as supervision, fund-raising, staff training, and continuing education.

Collectively, the mission, objectives, goals, and strategies of the organization reflect the organization's philosophy and values. In some cases, this philosophy is explicitly stated in the mission statement, while in others it is more implicitly suggested. Often, it is the agency's strategies that are most telling in this regard. Several agencies might serve identical populations and seek identical outcomes but have vastly different ideas and philosophies about how to reach those goals. Consequently, their strategies will be vastly different as well. For example, in the case of three programs serving teens with substance-abuse problems, each might pursue its goal of helping clients become drug free by different approaches. The first might encourage the youth toward spiritual transformation and reliance on a Higher Power. The second might use traditional psychotherapy in combination with carefully administered psychotropic medications. The third might use outdoor adventure camping, hard work, and group cooperation. Whether or not their philosophies are spelled out, clearly there are different ideas and values operating in each agency's understanding of the nature of the individual's problem as well as the strategies that are thought to be most effective in prompting change. The philosophy and values of an organization are compellingly powerful determinants of *how* the agency goes about accomplishing its goals. As you examine your agency's mission statement, identify the philosophy and values that guide the agency's work and shape its identity.

A Student's Reflections on the Agency's Mission Statement

As you can imagine I was less than thrilled when my supervisor suggested that I spend the first hour or two of the day just "getting familiar with our printed material." I cast my gaze upon a book, three inches thick, in a drab gray ringbinder. Some of it was dry policies and such, but the agency's mission statement really inspired me. I have since found myself thinking about that statement, particularly when I feel discouraged about my work. In the daily grind it's easy to lose sight of what we're really trying to accomplish here. I think if I were an employee here with my own office and desk, I would frame a copy of the mission statement and keep it at exactly eye level right over my desk.

Exercise 3.3

Explain, in your own words, the purpose or mission of your agency.

Describe the goals and objectives of your agency or, if you are in a very large organization, those of your department. If your agency's goals and objectives are not in written form, interview a few key people in your agency and inquire about the goals and objectives that are informally guiding the agency's efforts.

As you observe the agency's work, what specific strategies are being used to meet the agency's goals and objectives?

What do the agency's mission statement, objectives, goals, and/or strategies say or imply about its philosophy and/or values? How are the philosophy and values reflected in the agency's operations?

Based upon your observations thus far, what is this agency actually accomplishing for individuals? families? the community?

LEARNING ABOUT THE AGENCY'S ORGANIZATIONAL STRUCTURE

Your agency, unless it is very small or very new, probably has an organizational chart that identifies by job title the various positions in the agency and reflects who reports to whom within the system. The organizational chart conveys information about how the agency is organized to accomplish its work. Divisions, departments, and programs within the agency appear on the chart in such a way that the various units of the organization and how they relate to one another become clearer. Larger and more traditional organizations tend to follow a highly bureaucratic organizational structure, resulting in a chart that resembles a pyramid. The bureaucratic pyramid structure results in a vertical organizational chart, showing one person at the top, many people at the bottom, and an assortment of midlevel managers, coordinators, programs directors, and supervisors in between. Smaller organizations and those with less traditional, nonhierarchical management philosophies often have organizational charts that are more horizontal than vertical. These organizations use more peer supervision and collaborative methods of organizing and conducting their work.

Another equally important factor to consider in organizational structure is the nature of the formal organization as compared to the informal organization (Alle-Corliss & Alle-Corliss, 1998; Ehlers, Austin, & Prothero, 1976; Lauffer, 1984; Russo, 1993; Sweitzer & King, 1999). Although the agency has an official organizational structure that reflects who reports to whom, who works with whom, and who has power relative to whom, the actual working of the organization does not always conform to this official structure. The patterns of communication, relationships, and influence that emerge in the day-to-day work of the agency is referred to as its *informal* organizational structure.

The informal structure of the agency functions as a set of unspoken rules relating to every aspect of agency life, including how staff members relate to one another, how they relate to clients, how decisions are made, and so on. In general, organizations tend to function most effectively when the formal and informal organizational structures closely correspond to one another (Ehlers et al., 1976; Halley, Kopp, & Austin, 1998; Scott & Lynton, 1952).

As you become acclimated to the environment of your agency, you would be wise to observe both its formal and informal operations. Your ability to "fit in" with a particular agency depends in part on your ability to discern and follow the currents and crosscurrents of its formal and informal operations. The situation can at times leave a student feeling confused as to when to do what. Is it appropriate to talk to someone other than my supervisor about a given concern? Should I dress differently on days of particular events or meetings? Should I answer that phone or let it ring? By being a careful observer of the informal operations and interactions within your agency, you will probably be able to figure out how to conduct yourself in most situations. Another good rule of thumb is "when in doubt, ask." Your supervisor is the most obvious resource but certainly not the only one. Draw upon the relationships you are developing in the agency and raise questions as they arise. (Similar issues regarding agency functioning will be further discussed as aspects of agency culture in Chapter 6, "Dealing with Diversity.")

Exercise 3.4

Secure a copy of the organizational chart for your organization. If the agency does not have a current organizational chart, try your hand at developing one and have your supervisor critique it. Describe in your own words the formal organizational structure of your organization. Explain how this structure serves to help the agency accomplish its purpose.

Describe the informal structure of your agency, based upon your observation of it. How do the formal and informal organizational structures compare with one another? How is the agency's work affected by its informal structure?

LEARNING ABOUT AGENCY FUNDING

Most human service agencies today are funded by a patchwork of sources. Both public and private agencies generally receive revenues from multiple coffers. Depending upon its funding sources and its mission, your organization might be considered (1) a public agency, (2) a private nonprofit agency, or (3) a private for-profit agency.

Public agencies are funded primarily by tax revenues at the federal, state, and/or local level. Although these organizations may receive private donations and user fees, they do not rely heavily on these revenue sources. These programs exist due to legislative mandate and include such agencies as departments of social services, mental health centers, public health centers, and public schools, among others.

Private nonprofit organizations are funded primarily by private sources, such as religious organizations, private contributions, grants, and fund-raising proceeds. These organizations may receive some federal, state, or local revenues as grants for specific initiatives but not usually for ongoing operating expenses. In some cases, the government purchases services from private agencies. In such situations, the government establishes clear and specific contracts with these agencies to provide certain services to the public for which the government will pay (Ortiz, 1995). Many private nonprofit organizations have benefited from such government contracts for several years. In recent years, government contracts have become more common with private for-profit agencies as well (Gibelman, 1995). Also, depending upon the nature of the services provided, some agencies may qualify for reimbursement for their services from third-party payers, such as medical insurance companies and Medicaid, Medicare, or other government programs. Private nonprofits often receive grant money from private foundations and/or government sources as well. Some nonprofits may also charge small user-fees for specific services. Private nonprofits vary in size and complexity and include, for example, retirement centers which are funded by large religious organizations, large national organizations such as the American Red Cross, and smaller community organizations such as homeless shelters or crisis intervention agencies (Barker, 1996; Dobelstein, 1978).

Private for-profit organizations, sometimes referred to as *proprietary* social agencies, are initially funded by one or more individuals who invest their personal resources into the development of the organization. Examples of proprietary social agencies include private inpatient mental health and substance abuse facilities, educational institutions, nursing homes, camps, and so on (Barker, 1996). Once established, these organizations pursue a dual mission, to deliver services to a given population and to make a profit for their investors. Like some private nonprofit agencies, some for-profit organizations may also receive reimbursement for their services from state, federal, and/or local funds. Common examples are proprietary nursing homes that receive Medicare and Medicaid reimbursement (Garner, 1995) and for-profit day care centers that receive reimbursement from JOBS or "Workfare" programs for providing care for the children of their clients (Hagen, 1995). Additionally, some of these organizations also may qualify for third-party reimbursement from medical insurance companies. For-profit organizations also often have some clients who pay "out-of-pocket" for services. In general, proprietary organizations charge higher user fees than do public and private nonprofit organizations.

Depending on the type of agency in which you are working, the administration may be willing to provide a great deal of detail about its funding and budget or may be willing to discuss this only in more general terms. Agencies (both public and private) that receive government funds are required to be more open with this information due to issues of public accountability.

An agency's funding method has far-reaching implications for its operations. Some agencies must put a great deal of time into fund-raising, others have to work diligently to market their services, and still others must lobby and advocate for their budget needs through working with elected officials. It is difficult to make any generalizations about program quality based upon funding methods. Not-for-profit organizations are often seen as so underresourced that their services suffer. For-profit organizations are sometimes viewed suspiciously as the public questions the level of program quality that can emerge from a profit motive. Public agencies are often seen as being rigid, unresponsive, and overly bureaucratic. Each type of agency has its unique challenges, advantages, and disadvantages. It is my impression that there are excellent agencies within all three funding types and less effective agencies within all three types as well.

Exercise 3.5

Classify your agency using the three categories described above. In as much detail as possible, explain how your agency is funded. How are the daily workings of your agency influenced by the nature of its funding? What advantages and disadvantages does the funding method create for your agency?

LEARNING ABOUT THE AGENCY'S NETWORK

Because no one agency can serve all of the needs of its clients, human service agencies form an interdependent web of support and cooperation with one another (Halley et al., 1998). These relationships are generally informal, built from repeated interactions over long periods of time. At times, these relationships become especially cohesive and formalized as agencies pool their resources to accomplish some mutual goal, such as solving a community problem, providing a particular service or program, or even buying and sharing a piece of expensive equipment. Also, just as agencies cooperate to accomplish the work of serving the community, at times similar agencies may compete for resources, clients, or public recognition. Agencies serving the same target group may have difficulty establishing clear boundaries, delineating the identity of the programs as distinct from one another. Agencies may clash in their efforts to serve clients if those agencies have different philosophies of helping or different ways of viewing and conceptualizing client problems.

Much like the functioning of a family or any other social group, a certain amount of interagency conflict and competition is normal and predictable. Their presence should not be seen, in and of itself, as a problem. As in other groups, tension sometimes leads to productive growth and change. Agencies that work well together over time find ways of identifying the sources of conflict and resolving them productively (Brueggemann, 1996).

A Student's Reflections on Agency Networks

I began the day by attending the CAP (Community Alternatives Program) Council on Aging breakfast with my supervisor. The council is comprised of members of approximately 20 different organizations. Each of the representatives reported on their latest efforts to help the elderly population and the problems they were running into. Other representatives gave feedback on the various programs and how to solve the problems. One thing became obvious—all of these people are dedicated but they need more help. I learned a great deal about the various organizations and about the ways they cooperate in order to get more accomplished. I am impressed with the way these various organizations and professionals work together to make life better for the older adults in this community.

Exercise 3.6

To answer the questions in this exercise, you will probably find it helpful to talk with various staff members in your agency. Identify the particular agencies that work closely with your field site. When referrals are made by workers in your agency, where are these referrals most commonly made?

What agencies in the community refer clients to your agency?

Describe the nature of the relationship between your agency and the other agencies with which it interacts. In what ways do the agencies support one another? In what ways are conflict or competition present in these relationships, and how is this handled by the agencies involved?

How have the agency's relationships with other agencies developed and changed over time?

LEARNING ABOUT YOUR CLIENT GROUPS

Most agencies have a particular client group that they serve. Your agency might deal with the poor, preschool children, troubled adolescents, or the recreational needs of healthy older adults. Many agencies will serve more than one client group. As part of your field experience, you will want to make every effort to learn more about the groups served by your agency. In some cases, you will have studied the relevant populations in your course work. In others, you might have had some exposure through your own personal life or that of a friend, family member, or acquaintance. In any case you will want to be as knowledgeable as possible about the populations served. You can refresh your knowledge by reading some current professional literature relevant to serving these client groups. All human services workers and students are encouraged to read professional literature regularly as a method of expanding their knowledge about this and many other topics relevant to practice (Long & Doyle, 1999). Even if you have studied the population in question, you may have additional questions that you would like answered, or you might be curious about the latest research on particular issues or current methods used to address certain problems.

Exercise 3.7

Briefly identify and describe the client populations that your field agency serves.

List 5 to 10 questions that you currently have about these populations, their problems, and/or the human services interventions currently used to serve them.

Through your college library locate at least five current professional or academic journal articles and/or books that pertain to this target group. Try to find publications that deal with some of the questions you have raised. List below the complete bibliographical reference for each article and book, using APA style. This list should serve as a beginning reading list for your fieldwork. You may add to your reading list as new resources come to your attention. Bring selections from your reading list to your field site each day so that you can use any unscheduled time productively.

As you read from the reading list that you have created, look back from time to time at the questions you developed about your client populations. Try to answer them as you proceed through your reading. As you learn more, some questions will be answered and new questions will be raised. Discuss lingering unanswered questions with your field supervisor and/or with your faculty liaison. They may be able to direct you to additional resources to help fill the gaps in your knowledge. In some cases, it may be that your questions have not yet been answered in a systematic, scientific manner. In such situations, you may be in an excellent position as an intern to develop some tentative hypotheses and make preliminary observations on that topic. Practitioners with research skills can and do add significantly to our knowledge base in the human service field. Your fieldwork could provide a good opportunity to develop research interests that you might pursue in the future or even during your fieldwork with the approval and guidance of your field supervisor and faculty liaison.

LEARNING ABOUT THE CONTEXT OF YOUR ORGANIZATION

Every agency exists within a larger context. The realities of the local community as well as the larger legal, social, and political environment at the state and national levels can have a significant impact on life within the organization (Kerson, 1994). Such issues are important because they set certain boundaries on what the organization can and cannot do, which, of course, affects what you can and cannot experience as an intern in that setting. The nature of the local community and legal, social, and political issues can profoundly influence the functioning of the agency as well as staff morale, stress, and workload. Through considering these topics, you will gain a better understanding of the supports, tensions, and stresses that come into play as your agency performs its work against a backdrop of various forces.

Getting to Know the Community

Your first step in understanding the community context of your organization is simply identifying the community that your agency serves. This task might not be quite as straightforward as it seems at first glance because the community context for the organization is not necessarily synonymous with the town or county in which it resides. The agency's funding and mission are primary factors in drawing the parameters around the agency's community. For example, a state-funded school for children with behavioral problems may have the entire state as the community it serves, while a locally funded agency may serve only residents of a given county or city or even a given neighborhood. The shift in delivery systems from larger institutions to smaller, community-based organizations has created an increase in the number of agencies that serve fairly local populations as compared to the number of larger institutions that serve broader geographic areas (Mehr, 1998).

Capturing information about your agency's community will be a multifaceted task. Information about the community can be found in a multitude of places, but all of the information probably cannot be found in any one location. One of the best places to look for information about the community is in a recently conducted community needs assessment. Needs assessments of this type are conducted by numerous organizations including the United Way, the local hospital, the local community college, police departments, and/or county and city government planning offices. Some relevant information will also be available in U.S. Census reports. The public library or your college library may also be able to help you locate some of this information through its reference material and government documents. Also draw upon electronic sources for gathering information. State and local government Web sites as well as Web sites for the United States Bureau of Census are often excellent sources of information. Where no written, formal data exists, you might draw upon the impressions of long-standing community members. For example, a person who has worked in your agency for some time will probably hold some impressions about what particular geographic areas are heavily served by your agency, even if no hard data exists on the topic.

Exercise 3.8

Following are some key questions to answer as you get to know the community your agency serves:

What are the prevalent social problems and unserved needs in the community at the present time?

What is the general character of the community served by your agency? Urban, suburban, rural? How does this affect your agency's service delivery?

What are the major strengths or assets of the community? What are its major weaknesses or limitations?

What racial and ethnic groups are prevalent in the community? What special challenges are presented by cultural diversity in the area?

What relevant demographic information is available about the community? Gather statistical information about such characteristics as per capita income, family composition, household size, housing characteristics, unemployment rates, and age distribution within the community.

Are certain geographic areas of the community overrepresented or underrepresented in the agency's client population? If so, what ideas are there about why this might be the case? What methods might be used to more effectively reach and serve the underserved geographic areas?

What groups, if any, are overrepresented or underrepresented in the agency's client populations? What ideas are there about why this might be the case? What methods might be used to more successfully reach and serve underserved groups?

The information that you have gathered about your community may lead you to discover specific areas in your knowledge and skills that need development. For example, if your community includes ethnic or racial groups that you are not knowledgeable about, it would be helpful to read and learn more about these cultures before working with them. Similarly, being aware that you are working in a community with high unemployment can help prompt you to learn more about resources that might assist with this problem.

Understanding the Agency's Legal, Social, and Political Contexts

Kerson (1994) suggests that the laws that affect practice in a given setting and the prevailing social and political attitudes surrounding the work of a given organization strongly influence that organization's functioning. More formalized aspects of this context are sometimes collectively referred to as *social policy*. Social policy can be defined as "a set of carefully chosen guidelines that steer present and future decisions" (Halley et al., 1998, p. 97) regarding the well-being of people within a given society. Because it guides program and service development, existing social policy affects how human service workers intervene with clients and address human problems. Less formalized aspects of the social context for human service delivery include the attitudes, values, and opinions that are prevalent within a given environment.

Within the United States political system, there is a strong relationship between formation of formal social policy and these less formal dimensions of public opinion. Public opinion fuels political interest in certain social problems and dampens interest in others. Similarly, prevailing values and attitudes about certain social problems influence the shaping of social policy surrounding those problems. Thus, social policy and public opinion function as dynamic, interrelated forces that can strongly influence the human services delivery system and its various components.

Such issues are interesting and important factors to understand about your field site in that they form the larger backdrop for all of the agency's activities. Despite their importance, these issues might be particularly challenging to identify because they may not be obvious in the day-to-day routines of the agency. In examining the legal, social, and political context of your agency's work, you might think of the analogy of looking at your agency through a wide-angle lens rather than through a microscope. In doing so, you will think not so much about the agency's functioning in and of itself, but about the legal, social, and political forces at the local, state, and federal levels that affect practice and policy within the organization.

Some examples may be helpful in illustrating this level of agency context. On the federal level, major shifts in social policy have occurred in recent years regarding how to respond to the problem of poverty in our society. States are now required by the federal government to develop plans for reducing the number of individuals receiving Temporary Assistance for Needy Families, commonly referred to as "welfare." This policy, of course, affects the activities and climate within public welfare agencies, but it also affects a number of agencies of different types as well. For example, "reducing the welfare rolls" can increase client demands for other types of services, such as literacy programs, homeless shelters, child care, and crisis services. Such increased demands inevitably affect the day-to-day work of the organization as well as staff workloads. Prevailing values and attitudes that suggest that the poor should just "pull themselves up by their own bootstraps" can make it difficult for these agencies to garner the support and funding that they need in order to do their work effectively. In addition, the prevalence of such attitudes in the community and in society at large can create a climate in which the human services workers involved in serving the poor feel devalued and unappreciated, thus creating additional stress and discouragement within these organizations.

On the local level, the attitudes and values of a given community can strongly affect the types of human services initiatives that are and are not supported. For example, a community in which gay males and lesbians are more accepted might provide greater support for efforts to assist HIV patients or gay and lesbian parents than might a less accepting community. The demographic composition of the community clearly plays a role in this phenomenon. For example, in communities largely populated by older adults, the prevailing attitudes might be less supportive of services for children; while in communities dominated by young families, the opposite might be true.

At any given time, particular social problems seem to gain prominence in the national discussion. Certain attitudes can sometimes seem clearly dominant in these discussions, such as the strong public feelings currently about getting "tough on crime." In many situations, however, there seems to be more debate than consensus on certain issues as the attitudes and values represented in the discussion are more divergent. There have been lengthy and ongoing debates, for example, on such issues as the nature of sex education curriculum in the schools, contraception availability to minors, methods of curbing teen violence, methods of mainstreaming children and adults with disabilities into the community, humane treatment of the elderly, treatment of people with mental illness within managed care plans, and so on. Reading newspapers and news magazines will help you to stay abreast of important themes and ideas in the ongoing national discussion of various social problems. As a human service worker, these discussions can become very important to your work and to the clients you serve. Therefore, it is essential to pay attention to these issues, particularly as they relate to your agency and its clients.

A Student's Reflections on the Social Context of the Organization

Working in the area of teen pregnancy prevention is really tough and frustrating in a system that does not allow the discussion of contraception and requires that professionals only discuss abstinence with kids. In addition to this official policy, the school I am in is in a very rural part of the county where most of the students' families hold very conservative views. Even if the policy were different, there would still be a lot of pressure in this community to maintain the same system. As a result of this situation, it seems that the counselors' hands are often tied when it comes to helping the kids consider their options.

Exercise 3.9

What issues in the "national discussion" relate to your agency's work? What ideas, attitudes, and values about particular human and social problems influence the manner in which your agency goes about its work? In what ways is your agency's work encouraged and/or discouraged by these attitudes and values?

What state and/or federal policies and/or laws affect your agency's work? What philosophies, values, and attitudes are currently guiding policymaking as it pertains to your setting and the population it serves?

How do the values and attitudes of the local community impact your agency's work? How are these values and attitudes shaped by the community's demographic composition? For example, how might such factors as age, level of education, or political orientation of the residents of the community impact its values and attitudes and thus impact your agency's work?

CONCLUSION

This chapter has introduced you to a number of ideas and strategies for getting to know your agency, encouraging you to go beyond the more obvious features of the agency's identity. Because the agency has been shaped and continues to be shaped by a host of internal and external factors, these factors have been considered here. The agency's staff, mission, administrative structure, clients, and funding clearly play major parts in its identity and character. Also, the larger human services system in the community influences the agency's identity as it occupies a specific niche within that system. The nature of the community that the agency serves as well as legal, social, and political issues at the local, state, and national levels form a larger backdrop of forces that strongly influence life within human services organizations. The thorough understanding that you have developed of your organization through the discussion and exercises in this chapter will help you to understand more fully the various experiences you will have within that organization during the course of your fieldwork.

Your Reflections on Getting Acquainted with Your Field Agency

REFERENCES

Alle-Corliss, L., & Alle-Corliss, R. (1998). *Human services agencies: An orientation to fieldwork*. Pacific Grove, CA: Brooks/Cole.

Barker, R. (1996). *The social work dictionary*. Washington, DC: NASW Press.

Brueggemann, W. (1996). *The practice of macro social work*. Chicago: Nelson-Hall.

Dobelstein, A. (1978). Introduction: Social resources, human need, and the field of social work. In A. Fink (Ed.), *The field of social work* (pp. 3–21). New York: Holt, Rinehart and Winston.

Ehlers, W., Austin, W., & Prothero, J. (1976). *Administration for the human services: An introductory programmed text*. New York: Longman.

Garner, J. (1995). Long-term care. In R. Edwards & J. Hopps (Eds.), *Encyclopedia of social work* (pp. 1625–1634). Washington, DC: NASW Press.

Gibelman, M. (1995). Purchasing social services. In R. Edwards & J. Hopps (Eds.), *Encyclopedia of social work* (pp. 1998–2007). Washington, DC: NASW Press.

Hagen, J. (1995). JOBS program. In R. Edwards & J. Hopps (Eds.), *Encyclopedia of social work* (pp. 1546–1552). Washington, DC: NASW Press.

Halley, A., Kopp, J., & Austin, M. (1998). *Delivering human services: A learning approach to practice*. New York: Longman.

Kerson, T. (1994). Field instruction in social work settings: A framework for teaching. In T. Kerson (Ed.), *Field instruction in social work settings* (pp. 1–31). New York: Haworth Press.

Lauffer, A. (1984). *Understanding your social agency*. Beverly Hills, CA: Sage.

Long, L., & Doyle, M. (1999). Human services: Necessary skills and values. In H. Harris & D. Maloney (Eds.), *Human services: Contemporary issues and trends* (2nd. ed., pp. 131–142). Boston: Allyn & Bacon.

Mehr, J. (1998). *Human services: Concepts and interventions strategies*. Needham Heights, MA: Allyn & Bacon.

Neukrug, E. (1994). *Theory, practice, and trends in human services: An overview of an emerging profession*. Pacific Grove, CA: Brooks/Cole.

Ortiz, L. (1995). Sectarian agencies. In R. Edwards & J. Hopps (Eds.), *Encyclopedia of social work* (pp. 2109–2116). Washington, DC: NASW Press.

Russo, J. (1993). *Serving and surviving as a human-service worker*. Prospect Heights, IL: Waveland Press.

Scott, J. & Lynton, R. (1952). *The community factor in modern technology*. Paris: United Nations Educational, Scientific, and Cultural Organization.

Sweitzer, H., & King, M. (1999). *The successful internship: Transformation and empowerment*. Pacific Grove, CA: Brooks/Cole.

Thomlison, B., Rogers, G., Collins, D., & Grinnell, R. (1996). *The social work practicum: An access guide*. Itasca, IL: Peacock Publishers.

Weiner, M. (1990). *Human services management: Analysis and applications*. Belmont, CA: Wadsworth Publishing.

Chapter 4

Learning to Learn from Experience: The Integrative Processing Model*

After spending a week or two in their field agencies, most students begin to experience increasing responsibilities and a greater sense of independence. They often feel that they are starting to "test their wings" as human services workers. Also, many students begin to enjoy their fieldwork more, finding that they are busier and able to make more of a contribution to their organizations. Despite these positive developments, a word of caution may be necessary. Being busy does not ensure that learning is going on, nor does enjoying your work site and co-workers, nor does finding the work interesting. Learning from experience is a complex process that takes intentional effort, engaging and combining your intellect, behavior, and emotions. This chapter will help you understand more about the processes involved in learning by doing and will introduce to you a model for learning from experience that can enhance the quality of your thinking, learning, and performance in the field.

THE ROLE OF REFLECTION IN LEARNING FROM EXPERIENCE

Experience, in and of itself, does not necessarily teach very effectively. Your experiences in your fieldwork must be thought about, reflected upon, and analyzed to yield the greatest learning (Bogo & Vayda, 1995; Hutchings & Wurtzdorff, 1988; Kolb, 1984). You might think of experience without reflection as being somewhat like an unread book sitting or a shelf, or perhaps a book that you have quickly scanned but not carefully read, understood, or thought about.

A primary objective of your human services fieldwork is developing the ability to make meaningful connections between your practice experience and your knowledge such that your work is guided and informed by your knowledge of the field. If you are like many students, you might find this surprisingly difficult to do. A number of factors contribute to this difficulty. You might have studied relevant content and skills some time ago and, as a result, you may feel "rusty." As days become busier, sometimes even hectic, you may feel that "there isn't time to think" and only barely enough time to do

*Adapted and used with permission of *Human Service Education*. From Kiser, P. (1998). The integrative processing model: A framework for learning in the field experience. *Human Service Education 18*(1), 3–13.

what must be done. Another significant factor is the fact that you have done years of theoretical, intellectual work in the classroom, and now it feels good to be busy, active, and on the go. Falling into an unreflective way of working, however, is a trap that will severely limit what you learn as a student and your effectiveness as a professional. If you find that you are truly too busy to engage in reflection and planning, discuss this with your field supervisor. Your responsibilities may need to be changed so that you can do fewer things with higher quality.

Most of the time, students do have sufficient time for the integration of knowledge and theory with practice, but they need to set aside time for in-depth reflection and possibly even read and review relevant content. If you have saved your textbooks and notes from previous coursework, these will serve you well for such reflection and review during your internship. Additionally, students often need some guidance in *how* to do this integration of theory and practice. In essence, students generally need to learn how to learn from experience most effectively.

Case Example

George was a student intern in a day treatment program for troubled adolescents. Within the first week or two of his placement, staff members observed that George was overly controlling and critical toward the teens in the program. After about a week of observing such interactions and encouraging him to try other methods of interaction, staff members became frustrated with George's lack of progress and asked him simply to observe for the next week, not interacting with the clients at all.

Ironically, George had a 4.0 gradepoint average and could no doubt earn an "A" on any written test regarding theoretical information, including principles governing the development of positive helping relationships. His knowledge of theory, however, was of no use because he seemed to be unable to convert this knowledge into effective practice. During his week of "observation only," George's supervisor asked him to reflect on what he had learned in the classroom about helping relationships and to explore his own beliefs, assumptions, and feelings about the teens in the program. Although George was angry and defensive about the feedback he had gotten from the staff, he tried to respond to the questions his supervisor had raised by writing honestly in his journal each day, focusing on the issues assigned. After a few days, George and his supervisor reviewed what he had written and discussed it. In this discussion, George's supervisor was able to help him identify some erroneous assumptions, misinformation, and stereotypes that were contributing to his angry and controlling feelings toward the clients. George's supervisor was able to give him more accurate information about the teens in the program as well as about the staff. Although he continued to struggle some with his feelings, George was better able to use his academic knowledge in the subsequent weeks of his internship.

Exercise 4.1

What could George have done earlier in his internship to enable him to use his knowledge more effectively?

Reflect on your own internship. Try to identify any occasions in your own experience in which you might have used your knowledge more effectively. Also try to identify the obstacles that interfered with your ability to apply what you knew in these situations.

A Student's Reflections on Learning from Experience

I am so busy now that I seem to roll straight from one event to another all day long. My supervisor has also been busy so it is hard to even find the time to talk at any length. He is usually pretty reliable in making time for me though. I always think that I am learning as much as I possibly could, but then when we sit down together, he is always able to point out so much more that I didn't see before. It is the same with writing in my journal. The other night when I wrote in my journal I described a discussion I had observed between two of the workers on the unit. In the process of writing, I realized the two different theoretical perspectives (humanistic versus behavioral) that each of the workers was coming from in his point of view. I hadn't made the connection between the points they were making and the theories I had studied until then. It was neat to have this insight because it gave me more understanding of what they were talking about. It also made me proud of myself that I had put it all together without someone having to point it out to me.

AN OVERVIEW OF THE INTEGRATIVE PROCESSING MODEL

Much has been written about how students can best learn from experience. I propose in this chapter a six-step, cyclic process, which draws in part upon the strengths of various models that have been created about learning from experience. I refer to this model as the "Integrative Processing Model" (See Figure 4.1).

The model is described as "integrative" because it calls upon you, as a fieldwork student, to draw upon many components of yourself and your knowledge to extract

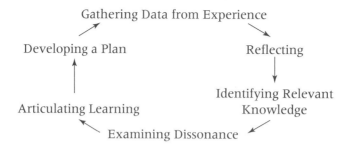

Figure 4.1: The Integrative Processing Model

maximum learning and meaning from your experiences. During your fieldwork, the Integrative Processing Model will serve as a tool to help you reflect upon and think through your field experiences carefully and systematically, integrating major components from your education, including knowledge of theory and content, self-awareness, and professional ethics. Your knowledge, behavior, attitudes, emotions, and values will all come into play as you process your experiences using this model. The Integrative Processing Model consists of the following steps: (1) gathering objective data from the concrete experience, (2) reflecting, (3) identifying relevant knowledge, (4) examining dissonance, (5) articulating learning, and (6) developing a plan.

The Integrative Processing Model is equally useful to those students working in direct services to clients and to those working in administration and other indirect services. You may use the model as a structure for written reflection and analysis of your experiences or as a purely cognitive process. The model provides an excellent structure for writing meaningful, constructive journal entries, and it is recommended that you use it as such. Write your responses to each step in the six-step model, processing several experiences each week in your journal. Writing your responses will help you to think through each step of the process thoroughly and carefully.

Because of the autonomy that students typically experience in their fieldwork, the Integrative Processing Model is designed for use as a somewhat self-guided process. It is assumed, however, that you will use it under the supervision of your field supervisor and/or your faculty supervisor. The field seminar provides an ideal opportunity to learn the model and practice using it as you analyze and discuss your field experiences with other students. The model also provides an excellent structure for discussing specific experiences with your supervisor during supervisory sessions. Over time, you will be able to use the model with greater autonomy and ease, applying it to everyday professional practice. With repeated practice you will internalize the model as a matter of habit, using it as an ongoing method for thinking about and learning from practice experiences. Thus, using the model routinely can enhance your learning, not only as a student during your field experience but subsequently as a practicing professional as well.

Each step of the Integrative Processing Model will be discussed in some detail in this chapter. Following the description of each step, you will be asked to apply that particular step of the model to a situation that you have encountered in your fieldwork. Before reading the following sections, select an experience that you have had during your field placement, which you can use as your practice example. It is most useful to select an experience that you found particularly interesting, challenging, troubling, or educational. If you find it difficult to implement a particular step in the model, it might be helpful to read the case examples later in this chapter. These examples will give you clear illustrations of how to apply each step in the model.

> ## A Student's Reflections on Processing Internship Experiences
>
> *One of the things that I have experienced during my internship is a shift in the way I think and work. I can summarize it best by saying that I have learned to put thinking and working together. In the beginning I was truly flying by the seat of my pants. Now I am much more focused, deliberate, and intentional in what I'm doing. Although there are always some "curve balls" that come along, I can usually explain why I handled a situation the way I did. I think about what I know (at least sometimes) and can actually use it to give me some ideas about what to do next or how to approach a situation. It is a good feeling to have all of these things start to come together.*

STEP 1: GATHERING OBJECTIVE DATA FROM CONCRETE EXPERIENCE

Models of learning from experience often suggest that learning begins with experience itself (Bogo & Vayda, 1995; Kolb, 1984). In your fieldwork, experience forms the basis for learning as each specific experience is an event from which learning must be extracted. The experience may be one in which you are an active participant or an observer. In subsequent steps of the Integrative Processing Model, this experience becomes the focus for reflection and application of knowledge. During the experience, as well as after the fact, you are gleaning information about the situation and about the behaviors, actions, and/or interventions of the various participants. Your ability to be an objective observer of experience is developed through this step of the learning process. In addition to your own observations, you might also use additional methods of observation. For example, you might arrange for your supervisor to observe an upcoming experience and provide an objective view of the situation. You might (with the permission of any other participants) audiotape or videotape the interaction. Writing process recordings in which you re-create the dialogue (both verbal and nonverbal) between yourself and others as precisely as possible is another good method of gathering objective data. An additional advantage of this approach is that it helps you to develop your own skills as a "participant-observer"—that is, training yourself both to engage actively in the interaction as well as to stand outside the interaction psychologically and watch it. Objective information, retrieved from the concrete experience by a variety of methods during Step 1, provides the focus for subsequent steps in the Integrative Processing Model.

Exercise 4.2

Select an experience to use throughout this chapter as your practice example. You will reflect on this experience repeatedly to practice the various steps of the Integrative Processing Model. As you think about the experience that you have selected as your practice example, consider the methods of identifying objective information discussed above. Using these methods, try to report objectively the important events, information, and facts from the experience you have selected and record those observations here.

STEP 2: REFLECTING

One of the most widely agreed upon precepts of learning from experience is the importance of reflection, as discussed earlier in this chapter. In Step 2 of the Integrative Processing Model you will engage in reflection by assessing your own personal reaction to the situation. You might think of this step as examining your "involvement of self" in the task at hand. How does this situation touch upon your own values? How does it relate to your personal history or similar experiences that you might have had? What thoughts and emotions does this situation trigger in you? What assumptions are you making about the situation? What assumptions are you making about the people involved in the experience, including yourself?

Another important component of reflection is evaluation of your own behavior in the concrete experience. As you examine your verbal and nonverbal behavior in the situation, what behaviors enhanced your effectiveness? What behaviors diminished your effectiveness?

Human services education emphasizes that professionals must develop and maintain a high degree of self-awareness in order to function effectively in the field. The reflection required in Step 2 helps to raise your awareness of the feelings, attitudes, behaviors, values, and assumptions that you are bringing to the particular experience.

Exercise 4.3

Using the preceding discussion as your guide, discuss your personal reaction to the situation. Be sure to consider the various components of reflection discussed previously in formulating your response.

STEP 3: IDENTIFYING RELEVANT THEORY AND KNOWLEDGE

Students engaged in fieldwork that is linked with professional preparation must make connections between what they have learned in the classroom and what they experience in the field (Argyris & Schon, 1974; Bogo & Vayda, 1995). In Step 3 of the Integrative Processing Model, you will identify theoretical, conceptual, and/or factual information that can shed light on your fieldwork experience. The information you have recorded in the previous steps of the model may constitute only a set of relatively meaningless, disjointed facts if examined outside the context of relevant theory and knowledge. Although previous classroom learning forms the foundation for Step 3 of the process, you might also need to engage in more extensive reading and research to expand your knowledge. For example, if you are working in an administrative role during your field experience, you might draw upon your existing knowledge of organizational theory and leadership theory but find that you need to do additional reading about models of supervision and strategic planning. If your fieldwork involves direct services to clients, you might draw upon your knowledge of theories regarding human behavior, the dynamics of the helping relationship, and the stages of the helping process but feel the need to do additional reading about a particular client's presenting problem or a particular issue in family dynamics.

Against the backdrop of relevant knowledge, certain information identified in the previous steps may rise to the foreground in its importance while other information might become relatively less significant. Some facts may begin to cluster together, bearing some relationship to one another, forming a more cohesive picture, pattern, or theme. The application of knowledge (whether theoretical, conceptual, or factual) provides an organizing focus, a lens through which you can view and make sense of the experience.

Exercise 4.4

Using this discussion about applying relevant knowledge as your guide, identify specific academic information that might be relevant to the situation you have selected. Consider theories and concepts that you have studied as well as any research or factual material that you are familiar with which relates to the situation.

STEP 4: EXAMINING DISSONANCE

Some authors advance the idea that dealing with dissonance is an important component of learning from experience (Argyris & Schon, 1974; Hutchings & Wurtzdorff, 1988). Dissonance is defined as "the lack of harmony, consistency, agreement" (*American Heritage Dictionary*, 1990, p. 409). Step 4 of the Integrative Processing Model suggests that, having examined the experience itself, your personal reactions, and relevant knowledge (Steps 1–3), you are now in a position to explore points of dissonance in the situation.

Dissonance can exist on a number of levels as you react to your experiences. You might experience intellectual dissonance as competing theories offer divergent points of view (Step 3) or as conflicting data arise out of the concrete experience (Step 1). You might experience dissonance between the espoused theories of the profession and your own personal views. You might find that your thoughts and feelings clash (e.g., "I know

I am supposed to be warm and empathic toward my clients, but I really don't like this person."). Conflicts might also occur between theory and practice ("I did what we discussed in class, but it didn't work") and between your thoughts and behavior ("I know what I am supposed to do, but I just can't do it right" or "I know what I am supposed to do, but I just can't make myself do it"). Following are some guiding questions that will help you to identify points of dissonance in your work:

- What, if anything, do I feel uncomfortable about in this situation?
- What disagreement is there between what I "should" do and what I "want" to do?
- What mismatch is there between what I "should" do and what I "must" do?
- What conflict is there between competing "shoulds" in the situation?
- What disagreement is there between my personal views of the situation and views offered by the theories and knowledge of the profession?
- What conflict is there between what I "know" and what I "do"? (Hutchings & Wurtzdorff, 1988)

Becoming clearly aware of dissonance is the first step in its resolution. Once recognized, such dissonance can often be resolved at the time it emerges. The values and ethical principles of the profession in many instances provide a useful framework for resolving dissonance. Values such as client self-determination, respect for the client's culture, recognition of client dignity and worth, appreciation of individual uniqueness, confidentiality, and nonmanipulative intervention are examples of useful guiding principles in reconciling conflicting points of view. Some points of dissonance, however, are more difficult to resolve. Particularly challenging are those situations that involve difficult ethical dilemmas. Such situations need careful consideration and consultation. (Chapter 7, "Developing Ethical Competence," deals with this topic in greater detail.) Other issues might present difficulty because they intersect with your strongest personal values or demand change in deeply ingrained assumptions or habits. These more troublesome issues may need to be discussed and considered for some time before resolution is achieved. If they present obstacles to your working effectively, these points of dissonance should become a central focus in your supervision.

It is also realistic to point out that at times the dissonance that you experience may not be reconcilable. You might find that rather than reconciling dissonance, you must instead learn to operate within it. In such instances, workers acquire the ability to tolerate ambiguity, to embrace competing values, to come to terms with their own personal limitations, and even to come to terms with the limitations of their profession.

Dissonance may not be involved in every experience, but it does occur frequently. Although the issues confronted in this step of the model are difficult and often uncomfortable, dissonance should be viewed as a necessary and desirable part of learning within the field experience. Through confronting and struggling with difficult issues, you will experience some of your most significant and meaningful learning. During this step, your total self is brought to bear upon the learning experience as your knowledge, skills, personal reactions, and values are wrestled with and brought to some degree of congruence with one another. Through this process genuine learning occurs, and professional maturity develops.

Exercise 4.5

Using this discussion about dissonance as your guide, identify points of dissonance that you experience(d) in the situation you have described in your practice example. Discuss the specific aspects of this dissonance that you need to work on resolving in order to gain greater clarity, comfort, and confidence in your work. Also identify any particular points of dissonance that might seem to be irreconcilable.

STEP 5: ARTICULATING LEARNING

Students often report with excitement that they are "learning so much" from their field-work, but when specifically asked what they have learned, they all too often fall silent. This scenario perhaps reflects the fact that, although students might indeed have learned a great deal, they have not thought carefully enough about this learning to be able to put it into words. Step 5, Articulating Learning, requires you to put your learning into words. Using words to explain and describe what you have learned will push you to conceptualize your learning. Through writing, a dim awareness can become clarified into a coherent statement. Once you have constructed this statement, the knowledge becomes more clearly your own. You then have greater command over this learning as a more tangible, concrete, and lasting "possession" that you can retrieve and use as needed.

The guiding question in this step of the model is straightforward, "What are the major lessons that I can take from this experience?" The lessons learned may have to do with skills that you developed, theoretical knowledge that you gained, insights that you developed into yourself or others, or deeper understanding that you acquired of an ethical principle. Whatever the lessons, remember that all knowledge is tentative rather than fixed and immutable. You will test the learning gained from a given experience in subsequent experiences. In doing so, the lesson may be reinforced, refined, revised, or refuted. Through this process, you will not only grow in your knowledge and skills but will also acquire the skills and habits of an active learner, equipped to continue learning from experience throughout your career.

A Student's Reflections on Articulating Learning

A new volunteer came in today to start her service hours for a course she is taking at the college. I was asked to spend some time with her and explain how things work around the agency. When I tried to explain certain things to her, I sometimes found that I wasn't really as clear about it in my own mind as I thought I was. I had to go ask my supervisor and other staff members questions just to get things straight. As a result, I learned some things from the process of orienting the volunteer. Until I had to put it into words I didn't even realize what I didn't know.

Exercise 4.6

Using the discussion about articulating learning as your guide, identify the major lessons that you are taking away from this experience.

STEP 6: DEVELOPING A PLAN

The final step, Developing a Plan, is a two-pronged step of the process in that it calls upon you to think through (1) how to proceed in your work (Bogo & Vayda, 1995) and (2) how to proceed in your own learning and development. Each of these topics will be considered in this section.

The work you have done in Steps 1–5 provides a solid foundation for making an informed choice as to how to proceed in your work. In some situations, the work of Steps 1–5 may lead you to a clear action path, but in many situations, a number of alternatives will be available. Step 6, Developing a Plan, consists largely of decision making through identifying, evaluating, and selecting from the various alternatives.

Ideas for alternative courses of action emerge in large part from Steps 3 and 5 of the Integrative Processing Model. Application of knowledge in Step 3, Identifying Relevant Knowledge, yields implications for practice, generating ideas for potential courses of action. Learning identified in Step 5, Articulating Learning, may also hold implications for what might be done in the next experience. Although Steps 3 and 5 are particularly helpful at this point, the work done in any or all of the steps of the Integrative Processing Model may offer ideas for additional options as well as support for or reservations about certain options. Additionally, common agency practices often suggest or reinforce certain options, while agency policy and protocol more clearly direct or even dictate the next step in some situations.

In Step 6, you will identify and weigh each alternative plan with consideration of the following: (1) What are the likely consequences of this plan? (2) What factors and forces support the selection of this plan and the likelihood of its effective implementation? (3) What factors and forces argue against the selection of this plan or present obstacles to its effective implementation?

Out of these considerations, you (in conjunction with your field supervisor) will select a specific plan for your next experience with the task or assignment. For the development of your own autonomy, it is best for you to function as independently as possible in developing your plan. Submitting to your supervisor written summaries of your experiences in which you work through each step of the Integrative Processing Model will be beneficial in this regard because it will ensure that you think independently, get feedback before you proceed with your plan, prepare adequately for your supervisory conferences, and use your supervision time in a focused, efficient manner. Due to time constraints alone, it is impossible to discuss thoroughly every experience that you have in your fieldwork with your supervisor. Written summaries will enable you and your supervisor to identify the most difficult situations for close review in your supervisory sessions, while less difficult situations might be quickly approved.

In addition to developing a plan for your future work, an equally important element in this step of the model is developing plans for your own learning. As you work through the steps of this model and make plans for subsequent work, you will more clearly recognize gaps in your knowledge and skills that need to be addressed. Developing a plan for your future work must include an assessment of whether you currently have the knowledge and skills required to implement that plan successfully or whether you need to acquire such knowledge and skills. A plan for your learning might include any number of activities, such as reading, observing another worker, attending a workshop or seminar, or doing a role play with your supervisor.

Once plans are developed for your next experience with the task or project, you are poised to begin the learning cycle again. Even in cases in which you will have no further immediate opportunities to work with the particular client, project, or task, you should still complete the final step by developing a plan. In doing so, you should look toward your next "concrete experience" as the next professional experience in which you will deal with a similar situation. Considering the question, "How might I handle things the next time I have such a responsibility?" will prove to be good preparation for this next experience whenever it should occur. Whether the next "concrete experience" is the next day or sometime in the future, thinking through your plan forms a basis for your activity in another professional experience. This next experience offers the opportunity to start the learning process again at Step 1, Gathering Objective Data from Concrete Experience.

Exercise 4.7

Continuing to focus on the situation that you have discussed in Steps 1–5, develop a plan as to how you will proceed in your work. Think about where you need to go from here in the immediate situation. If you do not anticipate further work in this immediate situation, think about how you might handle similar experiences in the future.

Finally, and very importantly, identify specific knowledge and skills that you need to gain in order to conduct your future work effectively. Develop a plan for gaining such knowledge and skills.

APPLYING THE INTEGRATIVE PROCESSING MODEL

The following examples of students using the Integrative Processing Model will help you understand more clearly how to use the model yourself. The examples will also give you some ideas about how to use the model as a structure for written reports.

Case Example 1: A Direct Service Situation

The following example is adapted from a student's encounter with an elderly client during her internship. The example serves to illustrate how the model helped her to think through the situation carefully and thoroughly, improving both the quality of her learning as well as the quality of the services that she was able to provide the client.

Step 1: Gathering Objective Data from Concrete Experience

> I visited Mrs. S, an 89-year-old widow, in her home. Mrs. S was using a kerosene heater to heat her small cinder-block home, and it was very hot inside. Her vision is quite poor, and she walks with a metal-frame walker. She does her cooking on a two-burner hot plate. Mrs. S. reported a lengthy story about her children stealing her land from her and wanting to put her in a nursing home. She repeatedly said that she needs a lawyer but can't afford one. She also said, however, that her children don't want her to go to a nursing home because it would take all her money and property. Mrs. S. was tearful and worried about her future. She worries about how she will take care of herself. She reported that she often goes for as much as a week or two without getting out of the house or having a visitor. I listened to Mrs. S. and empathized with her position. I told her I would find out what services might be available to help her stay in her home and that I would see if any legal help could be provided for her.

Step 2: Reflecting

I felt very sad for Mrs. S. She lives in a situation that I would not want one of my family members living in. I felt angry with her children for allowing her to live this way and not offering her more help. When I was talking with her, I bought into the story about her children taking her land because she was crying and so convincing. Thinking about it later, I realize that this might not be true. I do believe there is legitimate reason to be concerned for her safety due to her heating and cooking methods, combined with her physical problems. I have an impulse to try to rush in and fix everything here.

My grandfather lived with my family for the last 12 years of his life because he couldn't function very well on his own. I simply cannot imagine that none of Mrs. S.'s four children can help her out by having her move in with them or vice versa. This makes me really angry and makes me feel very sorry for Mrs. S.

Step 3: Identifying Relevant Theory and Knowledge

- Social support and stimulation are needed to help the elderly maintain optimal mental orientation as well as physical well-being.
- Elder abuse and exploitation are not uncommon and children are frequent perpetrators.
- The Department of Social Services must be notified if there is evidence of exploitation so that they can conduct an investigation of abuse and/or neglect.
- Dementia is also not uncommon among the old-old (those over 85) with paranoia sometimes being a symptom.
- Institutionalization can contribute to disorientation and physical decline as the individual has less control over his/her environment and possibly less opportunity or need to engage in self-care.

Step 4: Examining Dissonance

I am aware of the following points of dissonance:

- Mrs. S.'s story is confusing because she says her children want her in a nursing home and then says they don't want her in a nursing home.
- It is not clear whether the story about her property being stolen is true.
- I feel that it would be best for Mrs. S. to live with one of her children but she never indicated an interest in this.
- I am assuming that Mrs. S.'s children are unwilling to help her but I have not met them and am concluding this based only on her report.
- I perceive Mrs. S. to be in an unsafe living situation but she reported no particular problems living alone except for social isolation. No "close calls" were reported. Are my own middle-class lifestyle and values holding too much sway in my perceptions or is her situation really unsafe?
- It isn't clear to me where Mrs. S. wants to be. Does she want to stay where she is? Go to a nursing home? Live with someone? Some other option?
- I am uncomfortable with the fact that I told her that I would see if legal assistance might be available. Upon further thought, this might be premature.

Step 5: Articulating Learning

> I have learned that I must really have a soft spot for older people. As I look back at this experience, it is amazing to me how my emotions took the upper hand and how little I accomplished in this contact as a result. I think tears have a powerful effect on me, particularly when they come from an older person. I think my closeness to my own grandparents and my impulses to protect them come into play here. It is hard for me to see Mrs. S. as an adult. I will need to work on this as I go into my next contact with her as well as with other older adults.

Step 6: Developing a Plan

> Mrs. S. may be, but is not necessarily, a victim of exploitation. This needs to be explored further. I need to secure permission from Mrs. S. to meet with her children, perhaps in a family meeting. I am not sure I have the skills yet to carry this meeting off. It could be volatile and in any case will require some real skill and tact.
>
> Take a detailed history from Mrs. S. looking closely for any evidence that she has been in high-risk situations due to living independently. Also get the perspective of her children and neighbors on this if possible. I think I have the skills to do this and should have done so in the first meeting, but I got so wrapped up in her emotions — and mine!—that I didn't think of it. The potential obstacles might be the inaccessibility of her children and neighbors or her possible unwillingness for me to speak with them.
>
> Mrs. S. definitely needs more social stimulation, and she acknowledges her loneliness. Talk with Mrs. S. about possible friends, family, and other supports in her community that we might reactivate. Also see if she is interested in attending a senior center or congregate meal site. This should be easy. I can take some brochures from the senior center. She might be able to see the pictures and I can read them to her. If this is something she is interested in, she could probably get started very soon.
>
> Try to help Mrs. S. identify what she wants to do, where she wants to live. I can help Mrs. S. explore her options and again probably should have done this in the first contact. I believe it will be challenging due to her emotionality and her tendency to focus on the past. I will need to support her feelings but continue to focus and structure the interview on her options at the present time.
>
> In terms of my own learning and development, I think I should role play my next conversation with Mrs. S. There are several things that I need to accomplish in this meeting, and I know how easily I was sidetracked last time. I will ask my supervisor to role play this situation with me and give me any suggestions for conducting this interview more successfully. Also, for my own information as well as possibly for Mrs. S.'s, I need to research the local legal resources for indigent older adults. Whether or not this is called for in Mrs. S.'s situation, my conversation with her has made me aware that I really don't know how such situations are handled.

Case Example 2: An Administrative Services Situation

The following example illustrates how the Integrative Processing Model can be applied to administrative or indirect service experiences that students encounter in the field. As with the first example, the student's use of the model enhanced his learning as well as the quality of his work.

Step 1: Gathering Objective Data from Concrete Experience

As an intern in the hospital's human resource department, I was asked to organize a series of brief educational seminars on employee wellness. In a meeting with the human resources director, I learned a few facts about the need for the project. He informed me that the department had conducted employee surveys that revealed that two major concerns of employees were "too much stress" and "too little free time." The director had intended to implement some programs to deal with these concerns but did not have sufficient staff available to do so. Having an intern provided an opportunity to develop this long overdue program as well as an opportunity for me to develop my skills in organization and presentation. We decided that I would "kick off" the series with a session on stress management and that the lunch hour would be a good time for the session. Because I had done a presentation in one of my human services classes on this topic, I was well prepared, knowledgeable, and pretty confident.

I planned to spend most of the session presenting information to the group. Because there was not much time available, I needed to use it efficiently to cover the material. I started the session by explaining that the surveys they had completed indicated stress was a concern for them, and I asked them to identify some of their stressors. Almost immediately a number of participants went into lengthy tirades about the administration's personnel practices. A number of long-term employees had recently been "let go" in a drastic cost-cutting measure. Employees complained that those remaining were extremely overworked and fearful for their own jobs. One participant said, "I can't believe this stress-management thing is their response to that survey they did! No offense, but that's just throwing us a bone. We need some real change around here."

We got through about half of the program that I had planned. I did a lot of listening to and reflecting feelings, followed by efforts to redirect attention back to the topic. At the end, I told the group that I hoped they had picked up some useful information but that it mostly seemed that they needed a chance to ventilate. As they left, some participants thanked me and picked up copies of handouts that I had prepared. A couple of participants talked for a few more minutes about the difficult situation at the hospital.

Step 2: Reflecting

I felt really out of control throughout the so-called "presentation." From beginning to end, I felt uncertain as to what I should do and fearful that things were going to get even more out of control than they were. I vacillated between feeling angry at the participants for ruining my nicely planned presentation and feeling angry for them because of what they were having to put up with in their working conditions. I kept wanting to say, "Hey, listen, I'm only an intern," but I knew this wouldn't be very professional. Thank goodness I did remember how important it is to acknowledge and respect people's feelings. This one skill seemed to get me through the hour, although at the time it didn't seem like very much. I am pleased that I was able to maintain my cool and provide at least some support, but I was exhausted when the meeting was over. Also, I felt like I should do something for these people. I felt like marching into the administrator's office and setting him straight. I also found myself feeling angry at the human resources director for assigning me this project. Nobody prepared me for anything more than just a straightforward one-hour educational session. I keep feeling like I must have done something wrong for this to go so poorly. At the same time, however, I recognize that I did the best I could under the circumstances.

Step 3: Identifying Relevant Knowledge and Theory

- According to problem-solving theory, once a problem is identified, a number of steps should follow, including gathering information about the problem, generating a number of alternative solutions, and assessing and comparing alternatives. Selecting and implementing a solution should follow these steps.
- People who are highly skilled in their jobs and highly motivated to perform well function best with leaders who use a participatory style. Authoritarian or "telling" styles of leadership are most compatible with individuals who have low skills and low motivation.
- People who work in areas in which they experience high psychosocial demands and low control over their environment are at the highest risk of burnout. Frustration is one of the stages of developing burnout. Emotional outbursts are typical of this stage in the process.
- When two parties are in conflict with each other and a third party enters the situation, there is a risk of triangulation. Triangulation occurs when the third party gets "stuck in the middle," trying to solve the problem or "take up the banner" for one side of the conflict to challenge the opposite side.

Step 4: Recognizing Dissonance

This situation was so uncomfortable that the points of dissonance may be hard to enumerate, but here goes with the biggest issues:

- I am uncomfortable with the role I am in now. I feel that I need to try to do something helpful but I am not sure what to do. If I share the employees' concerns with the human resources director or other administrators, I am not sure what the consequences would be. Would the employees get in hot water? Would I get in hot water? I will need to discuss this session with the human resources director but I am not at all sure what to say.
- I am uncomfortable with the fact that I have agreed to lead more than one of these sessions during my internship. I am not sure this is a good idea now.
- I have always learned that the key to making effective presentations is being well prepared, knowledgeable about the topic, and well organized. It seems to me now that this isn't enough.

Step 5: Articulating Learning

I definitely learned more about what it means to be prepared for a presentation. Being prepared means a lot more than knowing your information, having an outline, and being well organized. I need to know as much as possible about my audience and their needs. Although I had thought I was well prepared, important planning had been completely omitted from the process. I should have known better than this, because I have studied the problem-solving process repeatedly. Now I think I finally get the picture. I also think I had made the mistaken assumption that administrative work was more straightforward and less emotionally sensitive than direct service. Wrong!

I also learned that I can I function pretty well on my feet. As I look back at the experience, I believe I handled it pretty well. I could have made the situation much worse by disregarding the feelings that were being expressed by the participants or by letting my own anger and frustration take control. I stayed tuned in to their feelings but also tried to stay task oriented.

Although I have experienced triangulation many times with friends and family members, I now see how easily it can come into play in the workplace, especially since I am the helping type who likes to fight for the "underdog." I know that I need to stay sensitive to this issue and make conscious decisions about when and how it is appropriate for me to get involved rather than be led purely by my instincts.

Step 6: Developing a Plan

Before leading another session (*if* I lead another session), I need to be sure that I clearly understand the nature of the employees' needs and concerns. I am not at all sure the problem-solving steps have really been worked through in this case. I will suggest to the human resources director that we need to gather more information about any identified problems before we try to put together a program to address them. Educational sessions may not be the best alternative. We need to rethink that assumption.

I need to try not to get "triangled" into the conflict between employees and management. I believe it would be best to acknowledge to the HR director that, as the survey pointed out, the employees were highly stressed and expressed a great deal of frustration in the meeting. I will explain that they seemed to feel more of a need to talk about the stress than to focus on stress-management methods at that particular time. I will suggest, as mentioned earlier, that it would be a good idea for the administration to gather more information about the nature of the employees' stress so that more precise interventions can be designed to address their needs.

If I do lead additional sessions in the future, I will use a more participatory style of leadership rather than a more formal presentation or "telling" style. This style is more appropriate for leading groups of highly skilled, highly motivated people and will allow for the expression of feelings that I now know are pretty intense. As a result I will need to prepare less information and be prepared to be flexible.

Also, as I interact with the employees, I will have a greater understanding of the situation, recognizing that the situation is one that frequently leads to burnout among helping professionals. This will help me not to personalize any reactions I might get, such as short tempers, frustration, or even apathy.

YOUR TURN TO PRACTICE INDEPENDENTLY

As you read about the Integrative Processing Model earlier in the chapter, you practiced each step of the model as it was introduced. Now that you have a better understanding of the model, you will need to practice using it every day in order to master the model and to learn as much as possible from your experiences.

Exercise 4.8

In order for you to practice using the model in a more independent and coherent manner, select another recent experience in your internship and process it using the Integrative Processing Model. Refer back to the discussion of each step as necessary to refresh your memory of what it involves and to ensure that you have thoroughly completed each step. Write your responses to each step in the space provided.

Step 1: Gathering Objective Data from Concrete Experience

CONCLUSION

Continue to use the Integrative Processing Model daily in your fieldwork. On most days, you will have at least a few, and possibly many, experiences that you can process using this model. Using the model to help you think through your experiences will create deeper and richer learning for you by ensuring that you engage in thorough reflection and critical thinking. The Integrative Processing Model provides an ideal structure for discussing significant events in your journal. As you develop more skill in using the model, you can become less structured in your approach, working through it mentally rather than on paper. Until then, continue to write your thoughts on each step. Even after you become more proficient in using the model, you will continue to find it helpful at times to write down your thoughts as you sort out more complex and challenging situations. With practice you will find that this process will begin to feel natural, comfortable, and helpful.

Your Reflections on the Integrative Processing Model

REFERENCES

Argyris, C., & Schon, D. (1974). *Theory in practice: Increasing professional effectiveness*. San Francisco, CA: Jossey-Bass.

American Heritage Dictionary. (1990). Boston: Houghton Mifflin.

Bogo, M., & Vayda, E. (1995). *The practice of field instruction in social work: Theory and process*. Buffalo: University of Toronto Press.

Hutchings, P., & Wutzdorff, A. (1988). Experiential learning across the curriculum: Assumptions and principles. In P. Hutchings & A. Wurtzdorff (Eds.), *Knowing and doing: Learning through experience* (pp. 5–19). San Francisco: Jossey-Bass.

Kolb, D. (1984). *Experiential learning: Experience as the source of learning and development*. Englewood Cliffs, NJ: Prentice-Hall.

Using Supervision

In the previous chapter you were introduced to a particular model for learning from experience. In that chapter you learned that experience itself and your ability to process that experience meaningfully are key resources in your learning. Throughout, however, Chapter 4 referred to the supervisor's role in helping you process your experiences. This chapter will further develop your understanding of the supervisory relationship and how to use it effectively.

A Student's Reflections on the Supervisory Relationship

I think my supervisor is a perfect one for me. I have worked for supervisors before who told me what to do, set up work schedules, and dished out a little praise or criticism as the situation called for. This supervisor is different. She really takes time with me and wants me to learn, but she doesn't hover over me. Sometimes she has thrown me into things that I didn't know how to deal with and I felt at loose ends. A few times I have even felt annoyed at her for that. As I look back on it, I think she has given me independence and support in about the right amounts. If she had waited until I felt 100% confident to expect me to handle things, I would never have done anything. I think it must be hard for supervisors to find the right balance between expecting enough and expecting too much.

UNDERSTANDING THE SUPERVISORY RELATIONSHIP

The field supervisor is an important key to student learning in any human services internship. You might think of your supervisor in a number of ways. Your supervisor may be thought of as a teacher who can impart to you new knowledge and skills. He may be thought of as an enabler, a supportive mentor who can help you to achieve your goals. She might be considered a broker and advocate, an established professional who can help you gain access to key experiences and people within the organization as well as within the community (Brashears, 1995). Ideally, supervisory relationships are based on trust and mutual understanding. Within the context of a trusting relationship, your supervisor becomes your primary and most immediate source of support while you are on the job (Shulman, 1995). Therefore, supervision should be seen as a valuable, desirable part of your working experience, both in your fieldwork and later in your career. Your willingness to draw upon your supervisor's support, knowledge, and expertise will enhance your learning and reduce your stress throughout your internship.

Despite the supportive nature of most supervisory relationships, many fieldwork students have not experienced professional supervision before and therefore often have mixed feelings about it. The prospect of supervision may be both reassuring and anxiety provoking. Although you want guidance and support, you might also have concerns about having your work scrutinized. You might have fears of not meeting your supervisor's expectations or of your work being criticized. Most fieldwork students feel more comfortable with supervision once they have formed a relationship with their supervisor. As is the case in other types of relationships, establishing good communication with your supervisor forms a solid foundation for your work together, so it is helpful to discuss with your supervisor your thoughts about supervision. It may also be helpful to recognize that although supervision starts during the human services training program, it is not unique to your role as a student (Neukrug, 1994). Supervision is a working relationship that you will have throughout your career. Wherever you work, whatever you do, you can be relatively sure that you will be accountable to a supervisor. Working effectively with supervisors is a key element in the growth and development of human services workers throughout their careers (Kaiser, 1997; Sadow, Ryder, Stein, & Geller, 1987).

Kerson (1994) points out that the supervisory relationship progresses through certain predictable developmental stages. She asserts that the relationship's development loosely parallels the eight stages of human development described by Eric Erikson. Hypothetically, following this model, the supervisory relationship begins with the establishment of trust, then moves into the establishment of student autonomy, followed by the development of student initiative, then student industry, and so on. This is a useful way in which to think about the supervisory relationship in that it holds a number of implications. First, it implies that the relationship develops over time. Secondly, it suggests that the student grows and becomes more independent within this relationship. Thirdly, it suggests the significance of the work that is done within the supervisory relationship, as the supervisor's role theoretically parallels that of the nurturing parent and the student's role parallels that of a young child striving to grow.

As this model suggests, learning to work productively with a supervisor and to use supervision effectively is an important professional skill that will help you to continue growing throughout your career. The exercises below will help you clarify your thoughts, feelings, and expectations as you enter this relationship.

Exercise 5.1

Think about how your relationship with your supervisor has developed thus far. Do you agree with Kerson's idea that the relationship loosely reflects the Eriksonian stages of human development? How has the development of your relationship with your supervisor been similar to and/or different from this model?

SUPERVISOR CHARACTERISTICS

Exercise 5.2

What characteristics do you consider to be ideal in a field supervisor? What hopes and expectations do you have about the type of supervision that you will receive in your internship? What particular concerns or worries do you have about your supervision? What specific requests, if any, would you like to make of your supervisor?

Thomlison et al. (1996, pp. 136–139) identify some traits of "ideal field supervisors." According to this source, the ideal field supervisor:

- is available
- is knowledgeable
- directs the student's learning
- has realistic expectations

Compare this list of characteristics to your own thoughts about supervision above. After examining the listed characteristics, you might have additional thoughts about what you need in a supervisor. It will be useful to share your thoughts about supervision with your supervisor. Keep in mind that any requests you might make are just that, requests. Your supervisor must decide how to conduct your supervision according your learning needs and his or her own supervisory style. Sensitive supervisors, however, will try to respect your preferences whenever possible, and they can do this best when you have clearly communicated with them about your hopes and needs.

A Student's Reflections on Supervisor Characteristics

I liked my supervisor on sight. She has a wonderful sense of humor, a quick mind, and a ready smile. She has a way of being friendly and businesslike at the same time. We went to lunch together where we discussed both of our goals for the fieldwork and how we should best go about achieving them. I appreciated her good sense in taking us away from the office as we were getting to know one another. She is a walking encyclopedia of information, but she shares it in a way that is natural and helpful, not superior. I can't help but learn if I spend much time with her.

Although most supervisory relationships develop fairly easily, sometimes there are special barriers to developing a positive relationship. Experiencing your supervisor as overly critical or unavailable to you are obvious examples of this. By the same token, having a supervisor who is far younger than you are or one who has a vastly different background can make forming the relationship more challenging. At times it seems that supervisors and their students can simply have very different personalities and personal styles, resulting in distance that may take a longer time to bridge in the relationship. Whatever the challenges, communication, patience, and flexibility will be key ingredients in ultimately building a satisfactory relationship. Although the relationship may develop more slowly, most students and their supervisors find ways to work together in a satisfactory manner. If you should find that your situation is particularly difficult and is creating an obstacle to your learning, you should discuss this with your faculty liaison.

A Student's Reflections on Supervision

My relationship with my supervisor started out a little uncomfortable. He is about 25 years old and has a master's degree and a few years of work experience. Here I am 45 years old, old enough to be his mother. In fact, I do have a son who is 21. I think my supervisor was a little intimidated by me at first. I was quite relieved when he brought up our age difference and asked me how I felt about it. It gave me a chance to tell him that I felt I could learn a lot from him regardless of his age, which seemed to help break the ice. It also made me feel good *that he said that he felt he could learn from me, too. I am glad that he doesn't see me like he would a 20-year-old, and I'm glad he had the courage to bring up the subject for discussion. The comfort level between us has been much improved ever since. Now that I am approaching the end of my internship, I can honestly say that I have learned a lot from him. I think the main things that we both brought to our relationship that made it work out so well were honesty, respect, and openness.*

Since supervision is a relationship based on two-way communication involving both the supervisor and the student, it is necessary to look at your own contribution to the relationship as well. We will now turn our attention to student characteristics, and how they influence the supervisory relationship.

STUDENT CHARACTERISTICS

Exercise 5.3

Imagine that you are a field instructor supervising a human services intern. What characteristics would you like to see in the student you are supervising?

How would you like the student to relate to you?

What specific traits and behaviors would you be concerned about if you saw them in your student?

Although you might feel that your supervisor has all the power and that you have none, rest assured that you too have a powerful role to play in shaping the supervisory relationship. The qualities that you bring to the relationship and the manner in which you communicate with and relate to your supervisor are critical to the quality of your relationship and how much you learn from supervision (Halley et al., 1998). Kaiser points out that supervisees have the power to "enhance or sabotage the supervisory process" (1997, p. 48). Although supervisors have the knowledge and expertise to provide a learning opportunity, the supervisee must be cooperative and open to what the supervisor has to offer.

Through working with many supervisors, it is my impression that certain characteristics are commonly considered to be desirable in interns because they are conducive to a positive learning experience. The "ideal intern" might be described as:

- accessible to the field instructor through openness to feedback and instruction
- eager to learn
- inquisitive and energetic
- knowledgeable on at least a basic level
- realistic about his or her own skills and knowledge
- willing to take risks in order to gain new skills and knowledge
- appropriately assertive, taking responsibility for his or her own learning and demonstrating initiative
- a good listener, observer, and communicator

This list of characteristics, of course, is not exhaustive. Royse et al. discuss an extensive list of characteristics that are desirable in human services interns. These characteristics include a desire to help others, an interest in the organization, maturity, promptness, reliability, a positive attitude, and good grammar, among others (1996, pp. 16–17). Interestingly, but not surprisingly, both students and supervisors want to work with someone who is accessible and ready to invest energy in the learning process.

Exercise 5.4

How does the "ideal student" you described in Exercise 5.3 compare with the characteristics discussed above? From reading the previous section, what particularly desirable student characteristics have you become more aware of? Which of the characteristics discussed in this section do you consider to be most and least descriptive of yourself?

The best way to find out about your particular field supervisor's expectations regarding students is to discuss the topic directly. In fact, you and your supervisor probably discussed these expectations at least briefly during your initial meeting prior to the beginning of the internship. At this point, if you do not feel that you are sufficiently clear about your supervisor's expectations of you, initiate some discussion about it. A good starting point for the discussion might be sharing the thoughts you have developed in Exercises 5.2 and 5.3, asking for your supervisor's reaction and input.

Clarifying expectations is an important step in establishing a positive relationship with your field supervisor. The relationship, however, is a means to an end and not an end in itself. As in working with clients, a good working relationship is a necessary, but not sufficient, condition for change and growth to occur. Within the context of this supervisory relationship, a good bit of work must occur throughout the internship.

A Student's Reflections on Working with a Supervisor

Talking with my supervisor helps me to process things. I generally tell my supervisor what is on my mind and what is in my heart. There is a very intense part of me and I am glad that I have a supervisor who can handle that, but I recognize that this isn't therapy. My supervisor is helpful in keeping me focused on the issues at hand but is patient in helping me wade through the various quandaries and emotions that come up along the way. Some parts of supervision, of course, are very cut and dried, just passing on information. But the best part of supervision is really talking through the issues that come up. I am glad that my supervisor is not too quick to tell me what to do, although I admit that sometimes I want him to do just that. He truly uses an empowerment model. He helps me figure out for myself what to do. By the same token, though, I know he won't just sit back and let me do something stupid or just plain wrong.

WORKING WITHIN THE SUPERVISORY RELATIONSHIP

The process of supervision requires work on the part of the student and the supervisor. You will probably interact with your supervisor in a number of different ways. You will ask a quick question in the hall; she will briefly check in with you on how a particular meeting went. Maybe you will have lunch together at times and issues related to your work will weave in and out of your lunch-time conversation. You might observe your supervisor's work, and he might observe yours. All of these are valuable parts of your learning experience, but none of them constitute formal supervision. Formal supervision occurs when you and your supervisor have a **planned contact** with one another, which both of you have **prepared for,** for the express purpose of thoroughly **discussing professional issues, planning subsequent interventions or projects,** and **generating feedback** (Alle-Corliss & Alle-Corliss, 1998; Bogo & Vayda, 1995; Chiaferi & Griffin, 1997; Thomlison et al., 1996; Wilson, 1981). To understand the work of supervision each of the key components in bold above need to be considered individually.

Supervision Is a Planned Contact

You and your field supervisor need to schedule a routine one-to-one meeting time, if at all possible. The norm in most field placements is about one hour per week of formal supervision time. This may vary, of course, depending upon the nature of the work and the issues at hand at any given time. Planned contact ensures that supervision time is a priority and not an activity that occurs "whenever things slow down." In most human services agencies, things almost never slow down. Another benefit of planned contact is that planned meetings allow sufficient opportunity for both you and your supervisor to prepare for supervision.

Exercise 5.5

In what manner are your contacts with your supervisor arranged? How regular and systematic are these contacts? To what extent do you need to work on establishing more planned, systematic contacts with your supervisor?

Supervision Is Prepared For

Your supervisory time is valuable and possibly difficult to arrange. You will want to use the time well. The best method for ensuring this is to prepare for the meeting by developing written summaries of important events that you need to discuss (as discussed in Chapter 4) and an agenda of your concerns and questions. Prior to each supervisory session, your preparation will involve taking a pulse on your current work. Reflect on the work you have been doing, the meetings and other interactions that you have observed, the decisions currently confronting you, the preparations you are making for future contacts with clients or for projects, and the interactions you have had with other staff members. Supervision obviously cannot focus on every experience that you have had and every pending event, so you will have to set some priorities. Questions to ask yourself as you plan for supervision are, "What issues in my work currently concern me most?" and "What are the time-sensitive issues that need my supervisor's attention *now*?"

Once you have decided upon a few priority items for supervision, your preparation continues by gathering the necessary data and organizing your thoughts and questions so that you are ready to present your concerns concisely. Plan your agenda for supervision with the awareness that your supervisor may also be planning to bring in items for discussion. Your supervisor might want to follow a certain project or case closely or might need to use some of the conference time for launching a new task or project in your learning plan. In order to coordinate the concerns that both of you are bringing to the meeting, it is helpful to your supervisor to receive your agenda a day or so in advance. If more time is needed, then this can be scheduled.

Exercise 5.6

What methods have you been using to prepare for supervision?

What ideas do you have about how you might prepare yourself and/or your supervisor more adequately for your time together?

Supervision Involves Discussing Professional Issues

As you and your supervisor bring items into the supervisory conference, a twofold purpose is being served: (1) the quality of your work for the agency is being monitored and enhanced, and (2) your knowledge, skills, attitudes, values, and behaviors as a developing professional are being encouraged, developed, shaped, and reinforced. Toward these ends, the professional issues discussed may cover a wide range of topics, including but not limited to the following:

- your work with clients
- projects that you have been assigned.
- your work and interactions with colleagues
- your understanding of and responsibilities to the agency
- professional values and ethics
- your personal reactions, feelings, attitudes, and biases as these relate to your work
- the supervisory relationship itself

As this list suggests, the potential topics for discussion are fairly wide ranging. It is important to clarify, however, that all discussions in supervision are in service to the two goals of supervision identified previously. Although supervision can, and often does, touch upon personal issues at times, supervision is not counseling or psychotherapy.

Exercise 5.7

Which of the issues discussed above do you tend to focus on most in your supervision? Which issues tend to be overlooked? What particular issues do you feel the need to spend more time on in supervision?

Supervision Includes Planning

Although there are numerous points of departure for discussion in supervision, all of them essentially lead to one overarching question in the end, "What are the implications of this discussion for improving my work in the future?" Attention is placed on such questions as, "What will I do in my next contact with this client?", "How will I relate to this particular staff member in the future?", "How will I handle myself in the next meeting?", and "How will I approach the next task in completing this project?"

You and your supervisor will examine various options for how to proceed with your work, select an appropriate plan, and discuss its implementation. As discussed in Chapter 4, an essential element in developing any plan for your future work is considering your ability to implement it. Based on the current level of your skills, can you carry out this plan independently? If you do not currently possess these skills, how can you acquire them? In some cases you might need to implement a given plan with another worker present for support, or you might observe another worker implement the plan as the next step in your learning. Through the process of making and implementing plans, your skills are continuously being reviewed and expanded.

Exercise 5.8

To what extent does your supervision focus on issues of planning for your future activities and learning within the internship? To what extent does it focus on ensuring that you have the information and skills you need to successfully carry out the plans you have put together? Are there steps that you need to take in order to include more planning in your supervisory time?

Supervision Generates Feedback

Getting feedback is one of the most crucial yet potentially difficult aspects of the supervisory relationship. Emphasis is placed on the issue of feedback in this chapter because your ability to both accept feedback and give it effectively are essential to your success in your fieldwork and in your career.

As a student of human services you might have learned about giving feedback to others effectively. One of the principles to remember in giving feedback is that feedback can be difficult to receive. Interestingly, receiving feedback of any kind (both positive and negative) can be among the most difficult and challenging tasks of supervision. Yet ongoing evaluation is an indispensable component of your learning and development, not just as an intern, but throughout your professional career. Your field experience provides a good opportunity to work on receiving feedback effectively.

Exercise 5.9

Think about some specific instances in which you have received positive feedback. (You may draw from experiences within your internship or from previous experiences.) In what instance(s) have you been able to accept and benefit from positive feedback? In what instance(s) did you tend to reject the feedback offered? What factors contributed to your acceptance or rejection of the feedback offered?

Think about some instances in which you have received negative feedback. (You may draw from your internship experience or from previous experiences.) In what instance(s) have you been able to accept and benefit from negative feedback? In what instance(s) did you reject the feedback offered? What factors contributed to your acceptance or rejection of the feedback?

As you receive feedback, both internal and external factors can influence your ability to accept it. The following internal issues, for example, affect how people receive feedback:

- Feedback is easier to receive when it matches a person's self-perception and is more difficult to receive when it conflicts with a person's self-perception.
- Internalizing feedback requires time and reflection on the part of the receiver.
- Feedback is easier to receive if the receiver trusts the source of the feedback (Patterson & Eisenberg, 1983).

A person's ability to accept feedback depends not only on internal factors but on external factors as well. How the feedback is delivered can strongly influence whether that feedback is accepted or rejected. Some common guidelines for giving feedback effectively are listed here. Effective feedback is:

- direct and specific, describing specific behaviors or actions.
- offered calmly.
- timely—that is, delivered soon after the experience.
- balanced, recognizing both strengths and weaknesses.
- offered, not forced, allowing the receiver to reflect and respond.
- helpful in generating ideas for alternative ways of doing things (Egan, 1994; Patterson & Eisenberg, 1983; Young, 1998).

Exercise 5.10

Review the situations that you discussed in Exercise 5.9. Examine these situations in light of the guidelines regarding giving feedback listed above, considering both internal and external factors. Which of these guidelines were observed and/or violated in these situations? How did this affect your ability to accept the feedback you were given in these situations?

Think about the feedback that you have received during your internship, particularly from your supervisor. In what ways has the feedback that you received generally conformed to and/or violated the guidelines previously suggested?

As your relationship progresses, you and your supervisor might discuss how your supervisory relationship is working. Some supervisors are eager for feedback from their students, asking for suggestions as to how the student's learning needs can best be met. Other supervisors may be less open to suggestions. The relationship works best when the student and the supervisor can give one another feedback about how the supervisory relationship is working. Each of you probably has some suggestions about how you can work together most effectively. If you should have such a conversation with your supervisor, remember that it is important to evaluate your contribution to the success of the supervisory relationship, not just evaluate your supervisor's role.

Exercise 5.11

Using the preceding guidelines regarding giving feedback, write some helpful feedback that you might share with your field supervisor at this point in your relationship, if he or she seems open to such a conversation. What are the particular strengths that your supervisor brings to your relationship?

How might your supervisor help you to get more out of supervision or more out of your fieldwork in general?

How comfortable are you with the idea of discussing this feedback with your supervisor? How open do you perceive your supervisor to be to receiving your feedback?

Evaluate your own role as a supervisee. What strengths have you brought to the supervisory relationship?

What have you done to make the most of your supervisory time?

What might you do to get more out of supervision?

CONCLUSION

Your supervision may be best thought of as a helping relationship in which a professional worker assists you in your learning. Supervision serves the dual purposes of monitoring the quality of your work and enhancing your professional development. For this relationship to work effectively, it must be understood as developing over time and be based on good communication. The supervisory relationship should encompass certain key elements such as planned contacts, preparation, discussion of professional issues, plans for future action, and feedback.

Your Reflections on Using Supervision

REFERENCES

Alle-Corliss, L., & Alle-Corliss, R. (1998). *Human service agencies: An orientation to fieldwork.* Pacific Grove, CA: Brooks/Cole.

Bogo, M., & Vayda, E. (1995). *The practice of field instruction in social work: Theory and process.* Buffalo: University of Toronto Press.

Brashears, F. (1995). Supervision as social work practice: A reconceptualization. *Social Work, 40,* 692–699.

Chiaferi, R., & Griffin, M. (1997). *Developing fieldwork skills: A guide for human services, counseling, and social work students.* Pacific Grove, CA: Brooks/Cole.

Egan, G. (1994). *The skilled helper: A problem management approach to helping.* Pacific Grove, CA: Brooks/Cole.

Halley, A., Kopp, J., & Austin, M. (1998). *Delivering human services: A learning approach.* New York: Longman.

Kaiser, T. (1997). *Supervisory relationships: Exploring the human element.* Pacific Grove, CA: Brooks/Cole.

Kerson, T. (1994). Field instruction in social work settings: A framework for teaching. In T. Kerson (Ed.), *Field instruction in social work settings* (pp. 1–32). New York: Haworth Press.

Neukrug, E. (1994). *Theory, practice, and trends in human services: An overview of an emerging profession.* Pacific Grove, CA: Brooks/Cole.

Patterson, L., & Eisenberg, S. (1983). *The counseling process.* Boston: Houghton Mifflin.

Royse, D., Dhooper, S., & Rompf, E. (1996). *Field instruction: A guide for social work students* (2nd ed.). White Plains, NY: Longman.

Sadow, D., Ryder, M., Stein, J., & Geller, M. (1987). Supervision of mental health students in the context of an educational milieu. *Human Services Education, 8*(2), 29–36.

Shulman, L. (1995). Supervision and consultation. In R. Edwards & J. Hopps (Eds.), *Encyclopedia of social work* (pp. 2373–2379). Washington, DC: NASW Press.

Simon, E. (1999). Field practicum: Standards, criteria, supervision, and evaluation. In H. Harris & D. Maloney (Eds.), *Human services: Contemporary issues and trends* (pp. 79–96). Boston: Allyn & Bacon.

Sweitzer, H., & King, M. (1999). *The successful internship: Transformation and empowerment.* Pacific Grove, CA: Brooks/Cole.

Thomlison, B., Rogers, G., Collins, D., & Grinnell, R. (1996). *The social work practicum: An access guide.* Itasca, IL: F.E. Peacock.

Wilson, S. (1981). *Field instruction.* New York: The Free Press.

Young, M. (1998). *Learning the art of helping: Building blocks and techniques.* Upper Saddle River, NJ: Prentice-Hall.

Dealing with Diversity

Students usually find that their field experiences give them the opportunity to interact with people very different from themselves, representing a range of cultural groups. This chapter will guide you through considering several important principles and skills in working with diverse groups. Exercises in the chapter will encourage you to apply these principles and skills to your fieldwork experiences.

A Student's Reflections on Diversity

One of my biggest surprises about this school is how diverse the student body is. First of all there is socioeconomic diversity. The school draws from parts of town with the wealthiest and the poorest children in the county. Also, the children come from all over the place. They range from white to African American to Hispanic to Laotian to Vietnamese. There are even some European children here—two students who recently moved here from Sweden and several children from five different German families who recently moved here after a local company was bought out by a German company.

AN OVERVIEW

According to the U.S. Census Bureau, the population of the United States is becoming more diverse. Based on the 1990 census, projections are that Black, Hispanic, Asian, and Native American segments of the population will grow significantly over the next 50 years. Whites are predicted to comprise a diminishing percentage of the total population over this time period. These demographic shifts are occurring due to differential birthrates among the various groups as well as immigration patterns. Additionally, those 65 years of age and older comprise one of the fastest growing groups in the country. It is projected that this group will represent more than one-fifth of the U.S. population by 2030 (U.S. Census Bureau, 1992). One of the greatest challenges that will face human services workers who are being trained today is that of working effectively with individuals, families, and groups from highly diverse backgrounds and cultures (Neukrug, 1994).

Due to the growing diversity of the U.S. population, the relationships between human services workers and their clients are becoming increasingly cross-cultural in nature. Cross-cultural helping relationships are those in which the client is from one cultural background and the worker is from another (Woodside & McClam, 1998). Such relationships present special challenges for the human services worker because, based on their different cultural backgrounds, the worker and client are likely to bring very different assumptions, values, beliefs, and communication patterns to the helping relationship.

UNDERSTANDING CONCEPTS RELATED TO DIVERSITY

For individuals within a given environment or group, acquiring the characteristics of that culture is generally adaptive, helping them to cope and meet their needs within that particular social setting. Culture has been defined as the "knowledge, language, values, customs, and material objects that are passed from person to person and from generation to generation in a human group or society" (Kendall, 1999, p. 597). Cultural characteristics include such variables as language and communication patterns (both verbal and nonverbal), attitudes, norms, values, relationship and kinship patterns, religious beliefs, and many others (Cox & Ephross, 1998). Culture can be a useful concept in understanding not only ethnic differences but also the more commonly experienced differences associated with gender (Bricker-Jenkins & Lockett, 1995) and socioeconomic class (Schlesinger & Devore, 1995).

Closely related to the concept of culture is that of ethnicity. Ethnic groups may be defined as a particular type of cultural group—that is, "a collection of people distinguished, by others or by themselves, primarily on the basis of cultural or nationality characteristics" (Kendall, 1999, p. 598). People within a given ethnic group generally perceive themselves, or are perceived by others, as sharing a common history and/or ancestry. Some writers also point out that the term is generally used to denote a group of people who maintain a shared culture within an environment in which a different culture is dominant (De Vos, 1975). Thus, Anglo-Americans, even though constituting a cultural group, would not be seen as an ethnic group in the United States due to their dominance within the larger culture.

Another important concept to keep in mind as you try to understand and work within a culturally diverse population is that of minority groups. A minority group may be defined as "a group whose members, because of physical or cultural characteristics, are disadvantaged and subjected to unequal treatment by the dominant group and who regard themselves as objects of collective discrimination" (Kendall, 1999, p. 600). As this definition suggests, minority group status is based more on the relative powerlessness of a group in the larger society rather than on the size of the group. Therefore, women because of their relative lack of power historically in relation to the dominant group in society might be considered a minority group, even though they comprise roughly half the population.

Although it is important to understand these important concepts related to culture, preparing yourself to work effectively with diverse populations, of course, requires far more than this. Those workers who become capable of working with diverse groups effectively must engage in the intentional building of competence and skills as cross-cultural helpers. Some basic cross-cultural competencies include gaining knowledge of other cultures through study and direct experience, developing a greater awareness of your own cultural make-up, decreasing your ethnocentrism, and increasing your respect of other traditions (Lum, 1999; Petrie, 1999). Further, some writers suggest that effective cross-cultural helpers can benefit from developing an understanding of biculturalism as a potential coping skill for clients who are struggling to adapt within cultures different from their own (Chau, 1991; de Anda, 1984; Lukes & Land, 1990; Schram & Mandell, 1994; Valentine, 1971). Finally, and most important, the human services professional must also be able to apply such knowledge effectively and skillfully in actual practice situations (Weaver, 1998). This chapter will help you to begin developing some of these skills as you think about issues of diversity related to your fieldwork experience.

EXPERIENCING DIVERSE GROUPS

Exercise 6.1

You have probably had the opportunity to observe and work with individuals from diverse groups during your field experience. This exercise will help you develop a clearer picture of this diversity. Using the prompts below as your starting point, describe the populations with which you are working during your field experience.

"Through my fieldwork I have the opportunity to observe or interact with individuals representing the following . . ."

socioeconomic group(s):

ethnic group(s):

minority group(s):

religions or belief systems:

sexual orientation(s) and genders:

region(s) of the country:

age group(s):

Review and reflect upon what you have just written. How do the people being served within your agency differ from one another?

How do the people being served in your organization differ from you?

INCREASING YOUR KNOWLEDGE ABOUT OTHER CULTURES

As you work with people from varied backgrounds, it is sometimes difficult to sort out those characteristics that are cultural traits as compared to those that are individual traits. Within all cultures there is a wide range of individual diversity. Groups that might superficially seem to be homogeneous are often surprisingly diverse. Hispanic or Latino people, for example, may appear to have similar backgrounds, but this is in fact a very diverse group. A Latino individual might have cultural roots in Cuba, Mexico, South America, Central America, Spain, or Puerto Rico. Each of these groups constitutes a distinct subgroup within the broader Latino ethnic group, and each has its own history and cultural identity (Longres, 1995).

To assume that all people within a given cultural group are alike or even similar is to stereotype. Although stereotyping is to be avoided, general information about a client's background and cultural identity is often useful. You cannot and should not assume that your clients conform to common characteristics within their culture, but having a general knowledge of their culture will help prepare you to understand perspectives and worldviews that might be different from your own. Having a better understanding of a client's culture will enable you to work with them in a more sensitive manner and possibly help you to understand the origins of certain behaviors that you might otherwise misinterpret. For these reasons, human services workers have a responsibility to learn about the cultures that their clients represent (Lum, 1999).

Exercise 6.2

Using the information generated in Exercise 6.1, identify the various groups or cultures that you need to learn more about in order to work knowledgeably and sensitively with your agency's clients.

Develop a reading list that will enable you to learn more about these groups and cultures during your field experience. Identify below at least five current professional books or journal articles that you will read to enhance your understanding of the populations with which you are working. List the full bibliographic reference, using American Psychologial Association (APA) style.

Out of your increased understanding of culture, you will be better prepared to recognize culturally learned behaviors as well as individual differences among your clients. Most important, you will be better equipped to develop helping relationships with your clients built upon acceptance and understanding of their cultural context. In addition to your reading, clients themselves are indispensable sources of information about their culture. Listening to clients with an attitude of openness and respect is perhaps the most important thing you can do to overcome cultural barriers in your helping relationships (Halley et al., 1998). The acceptance and respect that you convey through your interest in their unique backgrounds contribute immeasurably toward developing a positive and productive cross-cultural helping relationship.

As you read about various cultures and interact with clients from different cultural backgrounds, look for the strengths and assets within each culture (Lum, 1999). In many human services settings, it is easy to become focused exclusively on needs and deficiencies. Such an orientation may cause you to overlook the resourcefulness, resiliency, strength, and capabilities that are part of each culture. Recognizing cultural assets will help you to avoid negative stereotyping, especially toward groups typically thought of as "disadvantaged" (Nerney, 1998).

A Student's Reflections on Cultural Strengths

When I first started doing home visits I was very judgmental. I came away from different places thinking to myself, "Why don't they find a decent place for their kids to live? Why don't they clean the place up? Surely they could afford a bed!" I realize now that much of my reaction was based on culture shock and ethnocentrism. As I've gotten to know the families better I am noticing that I have different thoughts as I leave their homes. I often think about how resourceful many of these families are and how optimistic they can be in the face of great obstacles. I can see now how much the parents do care about their kids. In fact, they often have to work especially hard to keep their children safe in some of these situations. It is clear to me now that they value family far more than material things. I am embarrassed to say that as I think back to my first reactions, I believe that they reveal something about my own materialism and my love of my own creature comforts. These families have taught me that it is possible to have a good family life without these things.

In addition to focusing on strengths, it is useful as well to focus on the similarities and commonalties between yourself and those clients who, on the surface, seem to be quite different from you (Nerney, 1998). Although cultural competence requires you to focus on cultural differences, it is essential not to lose sight of those universal human qualities that transcend culture as well. Developing cultural competence means that you will develop the ability to recognize and empathize with universal human feelings, needs, and desires, despite the cultural barriers and challenges that various relationships might present. Understanding how these universal qualities tend to be expressed differently within different cultures will help you to recognize these qualities more readily in cross-cultural relationships. Recognizing the commonalities between yourself and the clients you serve will enable you to relate more genuinely, empathically, and effectively with clients from all backgrounds.

Exercise 6.3

Think about the various cultures that you have had an opportunity to work with during your internship. (You might find it helpful to review your work in Exercise 6.1 in order to see an overview of these groups). What particular strengths or assets can you identify within these various cultural groups?

Think about any particular individuals whom you have encountered during your field-work who seem to be very different from yourself. Also, think about cultures that you have encountered which seem to be very different from your own cultural group. Try to look past the more obvious differences to see the commonalties that are present. What commonalties are you able to observe between yourself and the other person? between your culture and other cultures you have experienced during your fieldwork?

DEVELOPING AWARENESS OF YOUR OWN CULTURE

All of us see the world through the "lens" of our own unique culture, although most of the time we are not aware of this lens. Becoming aware of the cultural lens through which you view the world is perhaps one of the most significant, and most difficult, aspects of working effectively with those from other cultures (Lum, 1999). In developing competence as a cross-cultural helper, exposure to persons from other cultures is necessary, not only because it enables you to learn about other cultures but also because it enables you to learn about your own culture and become more aware of your own cultural make-up. In fact, awareness of your own cultural make-up is often best developed through directly experiencing contrasting cultures (Alexson, 1993; Corey, Corey, & Callanan, 1998; Lee, 1991). One of the benefits of interacting directly with people of different cultural backgrounds is that it heightens your awareness of differences in attitudes and behaviors. Much of your cultural learning occurred so early in your life that it probably exists outside your awareness. Therefore, unless you experience cultural contrasts, you are likely to take your own culture for granted, seeing it as what is "right" or "natural." When your culture is all around you, it tends to be like the air you breathe—all important, but outside your awareness.

The importance of gaining cultural awareness through interacting with contrasting cultures is not limited to the dominant cultural group. According to Axelson, "a Black person in the presence of a group composed mostly of White members might be more aware of his or her Black ethnicity than in a group composed mostly of Black members" (1993, p. 32). Whatever your cultural heritage and identity, experiencing cultural contrast is an important element in developing greater awareness of your own culture. The diversity that you experience during your fieldwork probably offers multiple opportunities to experience this contrast.

Exercise 6.4

Below are listed some commonly cited cultural variables—that is, some important components of culture that vary from one group to another. Using these as a guide, describe what you have observed about your own cultural make-up in contrast to other cultures you have experienced in your fieldwork.

Sense of self, including assumptions about autonomy and responsibility to others

Communication and language, including verbal and nonverbal expression

Understanding of and relationship to authority

Use of space

Use of time

Food and customs around eating

Attitudes toward work

Definitions of success

Beliefs about how relationships should be conducted

The value placed on appearance and hygiene

Beliefs about education and the value of education

DECREASING ETHNOCENTRISM

The first steps in developing cross-cultural competence are learning about other cultures and becoming aware of your own cultural background, biases, stereotypes, and values (Neukrug, 1994). Through these initial steps you have developed an understanding that cultural traits are simply learned behaviors. Nevertheless, there is a tendency to see one's own culture as "right," "normal," or "natural" and other cultures as being "wrong," "abnormal," or "unnatural." This tendency is a manifestation of a phenomenon known as *ethnocentrism*. Ethnocentrism occurs when one assumes that one's own culture sets the standard against which other cultures should be evaluated, that one's own cultural patterns are best and right (Kendall, 1999).

In order to work effectively with people from different cultures, human services workers must strive to overcome ethnocentrism. By examining their assumptions and reactions as they interact with people very different from themselves, workers can become more aware of their ethnocentric reactions. Most, if not all, human services professionals have had thoughts and feelings in the course of their work that have brought them face to face with their own ethnocentrism. Ethnocentric reactions might be experienced as feelings of rejection, judgmental thoughts or feelings, or a sense of shock or even disgust in reaction to a client's behavior, attitudes, or lifestyle. These can be difficult experiences, especially for people who generally see themselves as accepting and nonjudgmental.

It can be difficult to recognize ethnocentrism in ourselves, just as it is difficult at times to recognize our own cultural make-up, as discussed earlier. Each person's cultural make-up might be thought of as a lens through which he or she views the world. This lens is an important element in ethnocentric reactions as it acts as a powerful filter on our perceptions, causing each of us to pick up on certain things in our environment and remain oblivious to others. This lens also "colors" how we interpret what we observe, assigning meanings that may or may not be accurate or appropriate to the reality of the situation. The following example will illustrate how different cultural lenses operated to give two students very different perceptions of and reactions to the same field setting. Each of their reactions might be considered an ethnocentric response.

Case Example: One Placement + Two Students = Cultural Insight

Gamal and Sherry were two human services students who were placed in the same child development center for their fieldwork. Gamal was from a working-class family that originated in the Middle East. Sherry was reared in an upper-middle-class family from the northeastern United States. Although the two students were in the same setting at the same time, they had vastly different reactions to and concerns about the children and families served by the child development center.

Gamal had requested a placement in a child development center, despite his family's strong views about the importance of traditional gender roles. In spite of Gamal's flexibility in thinking about his own roles, he found that he was uncomfortable when he saw little boys in the classroom play with dolls or with the play kitchen. In fact, he was so uncomfortable at times that he sometimes approached these children to initiate a different form of play. Gamal was also amazed by how much "stuff" all the children seemed to have—toys, books, games, records, videos. Not only did the center possess these items in vast supply, but the children brought even more of them from home. Gamal sometimes felt overwhelmed by all of it and frankly felt that it was very wasteful and extravagant.

Sherry, who was in this same setting, had not thought twice about the little boys playing with the dolls and the kitchen, nor had she really thought about the supplies, materials, and equipment within the center. All of this seemed quite natural to her. She was, however, very concerned about some of the children. She was upset to find that some children were left at the center for 10 to 11 hours per day, feeling that this was far too long for a child to be in a day-care center each day. She was also concerned to see some children wear the same clothes to the center two or three days in a row. She saw this as evidence that the children were not being properly cared for.

Gamal and Sherry worked under the supervision of the same field supervisor and talked with her independently about their observations and feelings. The supervisor noted their markedly different reactions to their experiences and recognized the real learning opportunity that was present in getting the two interns together for a discussion. They were amazed at one another's reactions to the center, and each was able to give the other a different perspective on the issues that concerned them. Most interesting of all, neither student recognized that he or she was having an ethnocentric response to the situation until they met together. Only through this discussion did the two students recognize how their own cultural make-up—that is, their own cultural lenses—colored their perceptions of their experiences in their fieldwork.

Experience 6.5

Think carefully about this discussion of ethnocentrism as well as the case example. In what ways might your cultural lens have colored your reactions to your field agency and/or the clients it serves?

Describe at least one experience or incident in your fieldwork in which you experienced an ethnocentric reaction. What thoughts and feelings did you have in that situation? As you describe the incident, explain the specific aspects of your own cultural experience that played into your reaction.

A high degree of self-awareness, maturity, and openness to personal growth are required to reduce ethnocentrism and become more accepting of other ways of being. In fact, several aspects of self-actualization identified by Maslow (1954) reflect the types of developmental shifts that are integral to overcoming ethnocentrism. Among other traits, Maslow describes the ideal, healthy, fully functioning person as one who is more accepting of self and others, more perceptive of reality, more independent of culture and environment, more resistant to cultural influences, and more capable of transcending cultural bias. All of these traits, according to Maslow, are central to achieving the highest levels of personal growth and development. Interestingly, these same traits will help you become more effective as a human services worker as well, enabling you to work more sensitively with culturally diverse groups.

As you strive to overcome ethnocentrism and develop respect for different cultures, it is helpful to remember the adaptive nature of culture for people in a given environment. It will be helpful to ask yourself, "How might this particular characteristic have helped this individual to thrive, cope, or survive within the environment in which he or she has lived?" Especially challenging for human services workers is the fact that at times their clients are hampered by their culturally learned behaviors. Although certain behaviors are adaptive and desirable within their own environments, the client may experience distress in certain situations because these same behaviors violate the norms of the dominant culture, creating conflict or stress. The client then is in the confusing situation in which a behavior that is constructive and positive in one setting is counterproductive in another setting. For example, the dominant Anglo culture of the United States places a high value on punctuality. One manifestation of this cultural value is that businesses and other organizations generally operate on a system of precise schedules. Some cultures, for example Appalachian mountain culture, manage time far more loosely (Axelson, 1993). Individuals from such cultures may experience severe disapproval if they bring these habits regarding time into the typical American workplace. Such situations, often referred to as "culture clashes," can seriously affect an individual's ability to experience satisfaction and success within a different culture. An individual's ability to hold a job and meet his or her own survival needs might be significantly hampered in such a situation.

UNDERSTANDING BICULTURALISM AS AN ADAPTIVE MECHANISM

As our country becomes more culturally diverse, human services workers must often help clients cope with culturally challenging situations in their own lives. The worker's role in these situations is, as in all helping encounters, to help the client explore and understand the nature of their difficulty and identify ways that it might be resolved. Some clients in such situations can feel quite threatened because they fear that they are being called upon to give up their own cultural identity in order to be successful. Other clients may discount the value of their own culture and see it only as an obstacle to be overcome if they are to be successful. The notion of biculturalism is sometimes helpful to individuals in such situations. Biculturalism occurs when an individual is able to adapt his or her own behavior to a particular culture as needed while retaining his or her primary cultural identity. The fully bicultural person is able to shift successfully back and forth between cultures, experiencing gratification of needs within, as well as making contributions to, both worlds (Chau, 1991; de Anda, 1984; Lukes & Land, 1990; Valentine, 1971; Van Den Bergh, 1991).

The major benefit of biculturalism as a coping strategy is that it respects the client's allegiance to and identification with his or her primary culture, recognizing the value of the primary culture to the client's identity and ability to cope within that social environment. At the same time, biculturalism allows room for the client to adapt to another culture as necessary in order to meet his or her needs within that environment as well. Within the framework of biculturalism, clients can experience success in new or different cultures while maintaining their primary cultural identity. Although this adaptation process might require a good bit of effort and energy, it tends to be less stressful than the alternatives of giving up the primary culture altogether or experiencing repeated frustrations and difficulties within the different cultural setting. The following case example illustrates such a situation.

Case Example: A Client Learns to Be Bicultural

Maria, a 33-year-old, single, Italian-American woman, grew up in an ethnic community in which her family operated a family business for many years. The norm within that community and within her family was an "all-for-one-and-one-for-all" mentality. Individual needs and wants were seen as not being as important as the needs of the group. There was a high value placed on hard work and each and every person carrying his or her own load without complaining. If a person was asked to help another, then compliance was automatic. Within this context, it was assumed that people were always available to help one another and that if people asked for help, they really needed it. Maria attended a local university while continuing to live at home and work in the family business. After graduation, she accepted a position in a large corporation and continued to live at home, working in the family business on weekends. Maria took her cultural learning about working within a group to her new position. She was bright, capable, hard-working, and a team player, all of which placed her in high regard within this corporate setting.

Despite its obvious strengths, Maria's cultural learning had left her with some significant gaps in her ability to cope in the corporate world. She had not learned to be assertive and set reasonable limits regarding her workload. In fact, within her culture, to do so would have been seen as selfish and irresponsible. Maria also had not learned how to take credit for her work. She repeatedly was in the position of doing all or part of another worker's project because he or she asked for help but then received no credit for this work as her co-worker in the end reaped the rewards. Maria did not feel that she could say "no" to her co-workers' requests for her assistance and did not have the assertiveness to even address what was going on with her co-workers or supervisor. She sought help when she had become very frustrated and angry about her situation.

A culturally attuned helper was able to help Maria see her situation in its cross-cultural context. Behaviors that worked for her in her ethnic community, based upon values that she prized, did not work as well in the corporate setting. Maria made some decisions about whether she wanted to stay in the corporate setting and evaluated what she would need to do to meet her own needs within that setting. Her human services worker helped her to see that she could choose, if she wished, to work toward acquiring behaviors more adaptive to the corporate culture (e.g., limit setting, assertiveness, and calling attention appropriately to her personal accomplishments), while maintaining certain behaviors of her primary culture. While recognizing the adaptations which she might make in order to find more satisfaction in her job, Maria also gained a clearer appreciation of the strengths of her primary culture, recognizing how its work ethic and emphasis on teamwork were helping her to meet the challenges of the corporate world.

Although it took some time and energy, Maria learned to be bicultural. Over time, she became able to shift between the corporate world and her ethnic community with relative ease. While she remained most comfortable within her primary culture, she felt that her adaptation to the corporate culture had given her an opportunity to grow in ways that she never would have done otherwise.

Like Maria, many clients from various ethnic groups, lower socioeconomic groups, or non-Western cultures often feel strongly bound to the consideration of group interests and less bound to self-interest in their decision-making (Ewalt & Mokuau, 1995; Petrie, 1999). This phenomenon has also been noted as a common one among women as compared to men (Gilligan, 1982). This issue is but one of many cultural patterns that can create challenges for individuals as they function in the dominant culture while maintaining their own cultural identity.

Exercise 6.6

Think about the clients with whom you have worked during your internship. Describe how one or more of these clients might be struggling with conflicting expectations and standards as they move between two cultures. How might the idea of biculturalism be useful in helping these individuals cope more successfully?

All clients will not be open to the idea of bicultural adaptation, and biculturalism certainly cannot be seen as a panacea. For those clients looking for ways to adapt within culturally different environments, biculturalism can be a useful idea.

UNDERSTANDING AGENCY CULTURE

As a human services worker, you might experience culture clashes at times as you take your own culturally learned habits and assumptions into environments where other cultural norms are dominant. In fact, entering your field agency might have presented a similar situation for you. Agencies and professions build cultures over time that can seem quite foreign to a newcomer. The term *corporate culture* recognizes that businesses and organizations function as distinct cultural entities, having their own clearly established systems of values, norms, assumptions, beliefs, expectations, behaviors, rituals, and so on (Holland, 1995; Weiner, 1990). Some agencies, for example, place a very high value on staff cohesion and camaraderie within the group, while others place a high value on individual effort and autonomy. Some agencies have very specific expectations regarding how employees should dress, while others have a far more laissez-faire attitude regarding dress. How staff members relate to each other, how they relate to clients, whether doors tend to be kept open or closed, whether lunch is a shared time within the group or a time for people to go their separate ways—all of these are examples of how agency culture expresses itself.

You entered your agency with a unique cultural identity of your own. In addition to the cultural heritage that you gained in your family and community, as a student you have also adapted to a particular campus or student culture. Given this mix of cultural realities, many students experience their first weeks in their field placement as a process similar to adapting within a new culture.

Experience 6.7

Review the first days in your field placement. What did you observe about the agency's culture (review the components of agency culture discussed earlier)?

How did the expectations of the agency culture differ from your usual experience in the "student culture" of your college or university? the culture of your family and community?

In what ways, if any, did you experience a clash between your behavior and the norms and expectations of the agency culture?

Although your agency culture might not have been drastically different from your personal culture or your campus culture, it is likely that you were able to identify at least at few areas in which you observed differences and possibly had to adapt your behavior. The clients with whom we work are often faced with similar, but often much more pronounced, culture clashes. An important skill in developing cross-cultural competence is gaining the ability to help clients navigate the cultural challenges that they face in our increasingly diverse world.

A Student's Reflections on Agency Culture and Student Culture

At first glance I didn't think there would be that much difference between life as a student and life in a human services agency, but by the end of the first week it was clear to me that there were some differences. The people here might look like me and my college professors, but the expectations were certainly different than on campus. First of all, because everyone had been so nice and friendly, I didn't realize that in some ways things would be so much more strict or formal. As with most things, I learned this the hard way. On the second day of my internship, we had a case conference that lasted for about one-and-a-half hours. After an hour or so, I got up and went to the bathroom and came back in. This would have been OK in most of my college classes, but after the meeting my supervisor explained to me that it wasn't "the norm" here. People stay in a meeting until it is finished unless they are called away for something really important, like an emergency. Some other things that I have come up against have been about things like phone manner and skirt lengths. I have noticed that my natural way of asking a person on the phone to wait is to say, "Hang on a minute." Nobody even has to tell me how awful this sounds or that it does not "fit" with this agency's way of doing things. When it *pops out of my mouth, it sounds horrible there, but it sounds fine to me at home. I am trying to develop a new habit of saying, "Would you please hold?" Also, recently a student volunteer came to work wearing a skirt that was too short. The volunteer coordinator asked her to go back home and change. You would see hundreds of skirts like it every day on campus. In fact, I have a skirt pretty much like it in my own closet and probably would have worn it to work eventually. At least on this one I got to learn from someone else's mistake!*

I've never worked in a professional environment before, and even though it doesn't bowl you over at first, there really does seem to be such a thing as a "professional culture." It is clear to me that I need to try to conform to this culture if I want to fit in.

CONCLUSION

Over the course of your career, issues of diversity are likely to become more pronounced in your work and in society in general. The principles discussed in this chapter will be helpful to you in working with this increasing diversity. The greatest barriers for human services workers as they form relationships with clients from different cultures tend to be a lack of cultural self-awareness, a lack of knowledge about culture, and ethnocentric reactions to cultural differences. Therefore, as a student, it is important that you learn as much as possible about the cultures of the various clients with whom you work and strive to become more aware of your own cultural make-up and reactions. Monitoring your reactions to various clients and situations can help you identify feelings of rejection or judgmental attitudes that might possibly be "red flags" indicating cultural bias or ethnocentrism that you can work to overcome. Once you have developed cultural competence, the cultural diversity that you experience in your work can be a real source of pleasure, adding interest, depth, and variety to your professional life.

Your Reflections on Dealing with Diversity

REFERENCES

Axelson, J. (1993). *Counseling and development in a multicultural society.* Pacific Grove, CA: Brooks/Cole.

Bricker-Jenkins, M., & Lockett, P. (1995). Women: Direct practice. In R. Edwards & J. Hopps (Eds.), *Encyclopedia of social work* (pp. 2529–2539). Washington, DC: NASW Press.

Chau, K. (1991). Introduction: Facilitating bicultural development and intercultural skills in ethnically heterogeneous groups. In K. Chau (Ed.), *Ethnicity and biculturalism: Emerging perspectives of social group work* (pp. 1–5). New York: Haworth Press.

Corey, G., Corey, M., & Callanan, P. (1998). *Issues and ethics in the helping professions* (5th ed.). Pacific Grove, CA: Brooks/Cole.

Cox, C., and Ephross, P. (1998). *Ethnicity and social work practice.* New York: Oxford University Press.

de Anda, D. (1984). Bicultural socialization: Factors affecting the minority experience. *Social Work, 29,* 101–107.

De Vos, G. (1975). Ethnic pluralism: Conflict and accommodation. In G. De Vos & L. Romanucci-Ross (Eds.), *Ethnic identity: Cultural continuities and change* (pp. 5–41). Palo Alto, CA: Mayfield.

Ewalt, P., & Mokuau, N. (1995). Self-determination from a Pacific perspective. *Social work, 40,* 168–175.

Gilligan, C. (1982). *In a different voice.* Cambridge, MA: Harvard University Press.

Halley, A., Kopp, J., & Austin, M. (1998). *Delivering human services: A learning approach to practice.* New York: Longman.

Holland, T. (1995). Organizations: Contexts for social services delivery. In R. Edwards & J. Hopps (Eds.), *Encyclopedia of social work* (pp. 1787–1794). Washington, DC: NASW Press.

Kendall, D. (1999). *Sociology in our times* (2nd ed.). New York: Wadsworth.

Lee, C. (1991). New approaches to diversity: Implications for multicultural counselor training and research. In C. Lee & B. Richardson (Eds.), *Multicultural issues in counseling: New approaches to diversity* (pp. 3–9). Alexandria, VA: American Association for Counseling and Development.

Longres, J. (1995). Hispanics overview. In R. Edwards & J. Hopps (Eds.), *Encyclopedia of social work* (pp. 1214–1221). Washington, DC: NASW Press.

Lukes, C., & Land, H. (1990). Biculturality and homosexuality. *Social Work, 35,* 155–161.

Lum, D. (1999). *Culturally competent practice.* Pacific Grove, CA: Brooks/Cole.

Maslow, A. (1954). *Motivation and personality.* New York: Harper and Row.

Nerney, B. (1998). Preparing interns to discover common ground. *NSEE Quarterly, 23*(4), 1, 16–22.

Neukrug, E. (1994). *Theory, practice, and trends in human services: An overview of an emerging profession.* Pacific Grove, CA: Brooks/Cole.

Petrie, R. (1999). Trends and challenges of cultural diversity. In H. Harris & D. Maloney (Eds.), *Human services: Contemporary issues and trends* (2nd ed., pp. 393–399). Boston: Allyn & Bacon.

Schlesinger, E., & Devore, W. (1995). Ethnic-sensitive practice. In R. Edwards & J. Hopps (Eds.), *Encyclopedia of social work* (pp. 902–908). Washington, DC: NASW Press.

Schram, B., & Mandell, B. (1994). *An introduction to human services: Policy and practice.* New York: Macmillan.

U.S. Bureau of Census. (1992). *Statistical abstract of the United States.* Washington, DC: Government Printing Office.

Valentine, C. (1971). Deficit, difference, and bicultural models of Afro-American behavior. *Harvard Educational Review, 41,* 137–159.

Van Den Bergh, N. (1991). Managing biculturalism at the workplace: A group approach. In K. Chau (Ed.), *Ethnicity and biculturalism: Emerging perspectives of social group work* (pp. 71–101). New York: Haworth Press.

Weaver, H. (1998). Indigenous people in a multicultural society: Unique issues for human services. *Social Work, 43,* 203–211.

Weiner, M. (1990). *Human services management: Analysis and applications* (2nd ed.). Belmont, CA: Wadsworth.

Woodside, M., & McClam, P. (1998). *An introduction to human services.* Pacific Grove, CA: Brooks/Cole.

Developing Ethical Competence

In recent years there has been intense interest in ethical issues within most professions. Issues of legal liability have become increasingly prevalent in the human services profession as well as in society in general. A number of factors are thought to have contributed to this increased interest in professional ethics and legal issues. Technological advances that have created new ethical dilemmas, difficult decisions regarding the allocation of scarce resources, and widely publicized ethical scandals and litigation within many professions have fueled the interest of professionals and laypeople alike regarding ethical practice (Reamer, 1995). Throughout your human service program, professional ethics has probably been strongly emphasized, and rightly so. Many authors have written at length about the significance of ethics in human service work. For example, Cournoyer writes, "Ethical responsibilities take precedence over theoretical knowledge, research findings, practice wisdom, agency policies, and, of course, your own personal values, preferences, and beliefs. . . . You must consistently use ethical decision-making skills in every aspect of your service to others" (1996, p. 53). Ethical decision-making is at the center of all that you will do in your fieldwork, all that you observe others do, and all that you will do in your career as a human service professional. Your fieldwork experience offers an ideal opportunity to develop and fine-tune your ethical skills.

A Student's Reflections on Ethical Issues

I have not really experienced any big ethical dilemmas or conflicts since I have been here, but sometimes I do think the agency could be a little more careful about some of the ethical standards. It is not at all unusual to go out to lunch with staff members who talk about clients all through lunch while sitting in a restaurant. Granted, names are not used, but still I think it could be a problem. I am always concerned that if people overhear, they might figure out who the workers are talking about, especially if it is a situation that has been in the newspaper or something.

AN OVERVIEW

Throughout each day, ethical issues are identifiable in each and every situation that you experience in the field. In many situations, however, you might not think about the ethical issues involved because the situation is so straightforward. For example, issues of privacy and confidentiality are involved whenever you close the door, talk with a client, and carefully handle the information shared. Issues of client respect are involved whenever you express acceptance rather than judgment toward a client. In these situations, your behavior is being directed by the ethics and values of the profession, although you may not consciously be thinking about this fact. At other times, the ethical issues in a situation take center stage, becoming painfully obvious because they are complex and not easily resolved. For example, at what point do you break confidentiality to report suspected child abuse? What do you do when you are angered or repulsed by a client's behavior? Your thoughts in such situations inevitably turn to ethical considerations as you search for a proper direction to take. Although some situations are inherently more ethically challenging than others, ethical action in all situations calls upon the worker's ability in the following areas:

1. The worker must have knowledge of the legal and ethical foundation of the human service profession.
2. The worker must understand the values and ideals that guide responsible human service practice.
3. The worker must be able to recognize the specific ethical, legal, and values issues involved in any given practice situation.

These skills are required in even the most straightforward situations in which there is not a values conflict or an ethical dilemma. When such conflicts and dilemmas do exist, the following additional skill is needed:

4. The worker must have the ability to weigh the conflicting ethical and value positions involved in a given situation and make decisions between them.

These four skills might be thought of as ethical competencies, each of which will be addressed in this chapter.

COMPETENCY 1: UNDERSTANDING THE ETHICAL AND LEGAL FOUNDATIONS OF THE PROFESSION

The foundation for ethical action rests in part on knowledge and understanding of the human service ethical code. (See the Appendix for the complete statement of *Ethical Standards of Human Service Professionals*.) In this section, you will read a summary of the *Ethical Standards*. As this summary reflects, the ethical code is broad and far-reaching, addressing professional behavior in a wide range of areas. In addition to professional ethics, this section will also deal with the legal standards that commonly affect practice in human services settings.

Ethical Standards of the Human Service Profession

Read the following summary of the standards with your field agency in mind. As you read, reflect upon situations that you have encountered during your fieldwork, as a participant or as an observer, in which one or more of these ethical issues came into play.

■ *The human service professional's responsibility to clients.* Clients have the right to information about the helping process, the right to receive or refuse services, the right to be treated with respect and dignity, the right to confidentiality, and the right to know about the limits of confidentiality. Workers are obligated to avoid dual worker-client relationships, to protect clients and others from harm, to maintain secure records, and to recognize client strengths.

- *The human service professional's responsibility to the community and society.* Professionals adhere to state, local, and federal laws and advocate for changes in those laws that conflict with clients' rights. Professionals stay informed of social issues affecting clients, engage in advocacy and mobilizing to satisfy unmet human needs, and advocate for the human rights of all people. Human services professionals are knowledgeable about their own culture and cultures within their community, respect cultural and individual differences, and seek training and expertise in working with culturally diverse populations. Workers act to protect the safety of others who might be harmed by clients' behavior and represent their own qualifications and services honestly.

- *The human service professional's responsibility to colleagues.* Professionals treat colleagues with respect, dealing with any conflicts directly with that colleague in effort to resolve the problem. Applying a similar principle, workers report ethical violations of colleagues, initially speaking to the colleague directly in an effort to resolve the problem. Workers avoid duplicating the efforts and relationship of another professional in working with a client and consult with other professionals as necessary to benefit the client. Consultations between professionals are kept confidential.

- *The human service professional's responsibility to the profession.* Professionals act with integrity, recognizing the limits of their expertise, seeking consultation, and making referrals as necessary. They continue to seek professional growth and learning for themselves throughout their careers as well as promote the continued development of the profession. Professionals promote cooperation among the various disciplines within human services.

- *The human service professional's responsibility to employers.* Professionals keep their commitments to their employer and participate in efforts to maintain positive working conditions and evaluate agency effectiveness.

- *The human service professional's responsibility to self.* Professionals strive to maintain personal qualities that are associated with effectiveness in the profession, foster their own self-awareness and personal growth, and commit themselves to ongoing learning and skill development (*Ethical Standards of Human Service Professionals,* 1996, pp. 11–17).

In view of the wide scope and rigor of these standards, Blair's statement regarding ethical practice is understandable when she says, "It would be unlikely that any human service worker could get through a whole month of working in the field without being able to provide some example of behavior observed that they felt was unethical, or perhaps breaking a law" (1996, p. 187). Human service work deals inevitably with important values and competing "shoulds" as well as real limitations on what can be done. For example, workers and agencies are expected to act in support of client well-being at all times but in reality often work with such limited resources that the quality of services suffers. Likewise, professionals are rightly expected to be empathic, nonjudgmental, and nonmanipulative toward their clients as well as direct and straightforward in addressing conflicts with colleagues. Yet workers, like all human beings, are fallible with emotions and impulses that are not always perfectly contained or managed. Within this mix of ideals with reality, it is safe to say that there are no perfect agencies and no perfect workers, though all must strive to maintain the highest possible ethical standards.

In addition to a solid knowledge of the ethical standards, it is wise to give some thought to which particular standards you are at greatest risk of violating within your own setting and within your own work. Each agency, depending upon the nature of the work it does, as well as its own inherent strengths and weaknesses, has its potential pitfalls. For example, in some agencies the intense nature of the demands upon the staff might increase the likelihood that conflict will develop between staff members; therefore, the standards having to do with colleague relationships might be particularly at risk. In another agency, tough decisions about confidentiality may be predictably at risk due to the population being served. For example, when working with potentially violent clients, issues regarding "duty to warn others" are likely to surface.

Exercise 7.1

Think about your own fieldwork site. What particular *ethical* standards are dealt with frequently within this setting? Identify the specific standards in question and explain why these standards are particularly pertinent and challenging within your setting.

Legal Issues and Standards of the Profession

Related to, but not identical with, ethical standards are legal issues that must also guide the decisions and actions of human service workers. Although professional organizations monitor and sanction their members' behavior in relation to the ethical standards of the profession, legal obligations carry even more weight. In the case of legal issues, the broader society sets certain expectations, duties, and obligations for professionals. Unfortunately, a clear, concise statement of these expectations does not exist. Legal standards change and evolve over time, emerging from court decisions, legislation, and regulations (Schram & Mandell, 1994). As a result of this dynamic and evolutionary quality, the legal standards impacting workers may not be completely clear because they are continuously being shaped (Cournoyer, 1996). Workers should be aware, nonetheless, that they work in an environment of not only ethical constraints but legal constraints as well. It is generally agreed that human services workers have the following legal obligations:

- to provide a reasonable standard of care
- to respect client privacy and maintain confidentiality
- to inform clients accurately and thoroughly about the services offered
- to report abuse, neglect, or exploitation of dependent persons
- to warn and attempt to protect persons who might be harmed by a client's behavior (Alle-Corliss & Alle-Corliss, 1998; Cournoyer, 1996; Everstein & Everstein, 1983).

Most of these legal standards are also addressed in the professional ethical standards, but the one area that is not directly addressed is that of "reasonable standard of care." The ethical standards do speak to issues of quality of care throughout in a general way, often identifying ideal standards of worker behavior. Although workers and agencies should strive toward the highest ideals in their conduct, unfortunately legal questions about standards of care often have to do with the lower limits of acceptable worker and agency performance, especially when litigation against workers and agencies is involved. Standard of care is perhaps the broadest of the legal issues, and one which clearly relates to each and every human services setting. It is morally and ethically right, as well as legally required, that all professionals and agencies provide a reasonable standard of care to the people whom they serve. But what exactly is a reasonable standard of care?

A reasonable standard of care relates to many areas of worker activity but generally is thought to include at least the following:

- knowledge about the population being served and the services being offered
- delivery of services and interventions based on sound theoretical principles
- reliability and availability of services to clients
- initiative and action on behalf of client and public safety
- adherence to ethical standards of the profession in relation to client care
- systematic, accurate, thorough, and timely documentation of client care

Exercise 7.2

Identify the legal issues and standards that are likely to be most relevant to your setting. What circumstances give these issues prominence in your setting?

As charges of malpractice become more common, human service professionals and the agencies in which they work must stay abreast of the legal issues that might emerge in their work. Malpractice charges may allege either professional misconduct or negligence on the part of a worker or agency. Therefore, acts of commission (what the worker did) as well as acts of omission (what the worker did not do) are open to such charges (Kurzman, 1995). Human services workers, of course, can reduce their risk of becoming involved in such charges by staying aware of the legal and ethical risk factors that are potentially present in their work and by consistently maintaining the highest standards of practice.

A Student's Reflections on Ethical and Values Issues

I know this isn't illegal but I certainly consider it unethical. There are a few workers here who talk to the children in a really hateful, angry tone of voice most of the time. I can honestly say that I have never heard two workers in particular say a nice word to the children. It's not my place to be correcting the paid staff members so I just bite my tongue. I can't believe that they would talk to adults this way. I believe that the human service values of treating people with respect, dignity, and warmth should apply to children, as much as they do to adults. I plan to ask my supervisor about this, but I don't want to seem critical. In the workers' defense, I have noticed that there is no training for new staff coming into the organization and very little in writing about how the staff should relate to the children. I believe that the workers at times fail the children by not giving them what they deserve, but I also believe the agency may be failing the workers as well by not giving them the training and support they need. Whatever the explanation for the problem, I believe that a higher level of care could be and should be provided for these children.

COMPETENCY 2: UNDERSTANDING THE VALUES AND IDEALS THAT SHOULD GUIDE RESPONSIBLE HUMAN SERVICES PRACTICE

In addition to having a knowledge of the ethical and legal standards of the profession, workers who strive to act ethically must also develop an understanding of the values and ideals of the profession that shape the larger context of ethical action. Certain worker characteristics may be thought of as professional values within the human service profession because there is a high degree of consensus among members of the profession that these characteristics are desirable. Some of the most commonly cited professional values are human dignity, respect, client self-determination, empathy, genuineness, positive regard, nonpossessive warmth, patience, self-awareness, pragmatism, noncontrolling interpersonal style, trustworthiness, and open-mindedness (Egan, 1994; Mehr, 1998; Neukrug, 1994; Patterson & Eisenberg, 1983; Schram & Mandell, 1994; Woodside & McClam, 1998). While some of these values are specifically mentioned within the *Ethical Standards of Human Service Professionals*, some are not. Yet the literature within the field clearly sets a context of expectations for competent professional behavior by consistently citing the traits previously noted, among others, as desirable. Each of these values potentially has direct bearing on the moment-by-moment choices made by human service professionals as they consider such questions as, "How will I respond to this individual?", "What action will I take in this situation?" or "What must I be careful not to do or say at this point?"

Exercise 7.3

Review the values listed previously as desirable worker traits. Expand upon this list by identifying additional commonly cited professional values not listed above, based upon your recall of your previous coursework.

Which of the professional values do you see as generally most descriptive of you?

Which of the professional values do you see as generally least descriptive of you?

Within your particular fieldwork setting, which of these values are you finding most difficult to live out in your daily decisions, behavior, and attitudes? Explain the aspects of your work that make these values particularly challenging.

COMPETENCY 3: RECOGNIZING THE ETHICAL, LEGAL, AND VALUES ISSUES INVOLVED IN ANY GIVEN PRACTICE SITUATION

Occasionally, interns will state that they have encountered no ethical issues during their internships. What students usually mean by this is that they have not encountered any gross ethical violations. The student might have a good knowledge of professional ethics, legal standards, and values but not be fully attuned to their relevance in everyday situations. An important objective of the fieldwork experience is that students develop ethical sensitivity—that is, the ability to recognize the ethical, legal, and values issues involved in their day-to-day work. Ethical sensitivity, as compared to ethical knowledge, goes beyond just knowing the ethical standards or recognizing gross ethical violations and extends into the ability to recognize more subtle ethical, legal, and values issues that operate within ordinary, everyday practice situations.

Human services professionals are expected to be sufficiently aware and reflective about their work to know the issues inherent in a given situation before they get pulled unknowingly into questionable behavior or circumstances. As a student, you are particularly urged to take time for careful reflection each day. Using the Integrative Processing Model (discussed in Chapter 4) will help to ensure that you are thinking through your experiences carefully enough to identify feelings of dissonance and to recognize the ethical, legal, and values issues that are present within them. For practice, it may be helpful to look at a few examples of difficult situations that other students have experienced and identify the issues involved.

Case Example 1

Anna was a human services intern at a crisis center for abused children. The center consisted of several components, including a group home, a victim advocacy program, and a community education program. Each program had its own coordinator who reported to the agency director. Anna's field supervisor was the agency director, Tyrone. Tyrone suggested that Anna spend time in each of the agency's programs, shadowing and accepting assignments from each of the coordinators. This was agreeable with Anna as it conformed to her own hopes of getting a broad exposure to the full range of services provided within the agency.

Within a few days, it became clear to Anna that there was a significant amount of tension between two of the program coordinators. Julie, the shelter coordinator, complained to Anna about Jeff, the community education coordinator, and Jeff complained to Anna about Julie. Jeff enjoyed spending time at the shelter and felt that the children needed some attention from a caring male as well as a good male role model. Julie felt that Jeff was intrusive and should focus on his own job rather than spending time at the shelter. Jeff felt that Julie was overly controlling, overly possessive, and overly protective with matters concerning life at the shelter.

For the first couple of weeks, Anna listened to both Julie and Jeff, making minimal responses and trying not to get involved in their conflict. Soon, however, Anna began to experience some of the same feelings toward Julie that Jeff had described. When Anna tried to get involved with the children at the shelter, she felt that Julie often assigned her something else to do to put distance between Anna and the children. She would ask Anna to run errands or get supplies ready for an upcoming meeting instead of allowing her to interact with the children.

Anna shared with Jeff her feelings and observations about Julie's behavior toward her. Soon Anna and Jeff were talking daily about the topic, trying to think up ways to crash through what they perceived as Julie's excessive and unhealthy control of life at the shelter. One day as they were talking in the crisis center office, Julie suddenly stormed into the room, furious at both Jeff and Anna. While they thought that Julie was at the shelter as usual, she had been doing some work in the room next door and had overheard Anna's and Jeff's remarks, criticizing and ridiculing her. She furiously reported that she would be speaking to Tyrone about the conversation she had overheard, and in the meantime she did not want either of them setting foot in the shelter.

Upset to the point of tears, Anna felt angry and misunderstood. She felt that she herself was the "victim." After all, it was she who had been assigned and had completed all of the "dirty work" from Julie, and now Julie had the nerve to attack her. On another level, however, Anna felt very uncomfortable with her own part in the problem. She knew that she had contributed to the problem and should have made some different choices about her own behavior along the way. Unfortunately, Anna had not recognized previously that there were ethical issues involved in this situation. Her awareness of this came only after her unpleasant confrontation with Julie prompted extensive reflection.

Exercise 7.4

What ethical, legal, or value issues were involved in this situation?

What might have happened differently if Anna had recognized the issues in this situation and her own obligations earlier?

What options were open to Anna early in the placement when she first began to experience problems with Julie?

Which option(s) would have been most desirable for her to implement?

What options are available to Anna at this point, following the confrontation? Which option(s) would you select if you were in Anna's position?

Case Example 2

David, a senior human service major, was excited about his placement in the local probation and parole office for his internship. He had the opportunity to work with and observe several staff members and was then given a small caseload of his own to monitor and work with. His faculty supervisor noticed in reading David's journal that there was a great deal of variation in his communication with his clients. Some clients he spoke with often and conducted professional helping interviews involving important exploration, decision-making, and goal setting. Other clients he spoke with rarely and did little more than lecture and give commands. The faculty member spoke with David about this during a site visit, and David responded with the following comment, "You see, one thing that I've learned in this field placement is that for some of these people, it is not even worth trying. They don't care about their own life. They're not making any effort, so why should the probation officer make an effort? You can just tell which ones are worth the trouble and which ones aren't. I've talked with several officers about it, and they all agree. You're just wasting your time trying to help a lot of these people."

Exercise 7.5

David does not seem to be aware of the professional issues involved in this situation. What ethical, legal, or values issues are present in David's situation?

If you had the opportunity to respond to David, what might you say?

Assuming that David has acquired his attitudes and behaviors from professional role models within the agency, how might he address these issues day to day in his work as an ethically sensitive helper?

Case Example 3

Keisha, a human services student, was performing her human services practicum at a child development center in a relatively affluent part of town. The center served mostly middle-class, professional, or two-career couples but also served a small group of children whose care was paid for by the Department of Social Services (DSS) as part of the Workfare program. As Keisha started her fieldwork, the center was embroiled in a child abuse and neglect allegation involving one of the DSS children. Staff members had repeatedly noticed marks on the child that resembled a belt buckle. The decision was made to report their observations to DSS. The staff explained to Keisha the laws that required them to report suspected child abuse or neglect. They emphasized that reporting was mandatory and that they had no choice, legally, ethically, or morally, about making the report.

Within a few weeks, Keisha noticed similar marks on another child while supervising a bathroom break. She reported her observation to the teacher in her classroom. The teacher said that she had noticed similar marks on the child before but felt sure that this was not an abusive situation because this child came from such a "wonderful family." She explained that the parents were well known in the community. The father was a successful, well-liked attorney, and the mother was enrolled in a local college finishing her degree. They lived in a beautiful neighborhood, and both were pleasant, responsible people. The subject was then dropped.

Keisha continued to see marks on the child from time to time and always reported her observations to the teacher. She became more and more concerned when nothing was done and felt very confused about what her own role should be in "going over the teacher's head" about it. After discussing the situation in field seminar, Keisha decided to talk with the teacher directly but calmly about the seriousness of her concerns. In this conversation she even brought up the contrast in how this child was being dealt with in comparison to the DSS child. Despite her frustration, she was careful not to become accusatory toward the teacher or the agency. The teacher responded well to the conversation and suggested that they meet together with the director of the center to seek some guidance on the issue. The director asked both Keisha and the teacher to begin reporting their observations directly to her. She explained that she would document and verify these reports with her own follow-up observations of the child. She explained that she felt that they could not make a report at that time because "nothing was documented."

Exercise 7.6

What ethical, legal, or values issues were involved in this situation?

As a student in the agency, did Keisha's options for dealing with this situation differ from those of a staff member? Why or why not?

What is your evaluation of Keisha's handling of the situation? What are the positive aspects? What are the negative aspects?

Like the students in the examples above, all human services agencies and workers confront ethical, legal, and values issues on a daily basis. Ethically sensitive workers are aware of and are able to identify the many issues involved in everyday practice situations.

Exercise 7.7

Identify at least one situation that you have either dealt with directly or been aware of in your field placement in which the ethical, legal, or values issues were particularly interesting or challenging. Describe the situation clearly.

Identify the issues involved in the situation, categorizing them accurately as legal, ethical, and/or values issues.

COMPETENCY 4: WEIGHING CONFLICTING ETHICAL AND VALUES POSITIONS IN A GIVEN SITUATION AND MAKING DECISIONS BETWEEN THEM

Many practice situations involve a number of ethical, legal, and/or values issues, each potentially implying a different course of action for the worker. These are among the most challenging situations confronted by human services workers. Some method must be used for sorting out the various issues involved and deciding which should take highest priority in the worker's decision-making and interventions.

Loewenberg and Dolgoff (1992) suggest that a hierarchy exists among various ethical principles and that workers should give higher priority to some principles and lower priority to others when ethical principles compete. They refer to this hierarchy as the Ethical Principles Screen. The Ethical Principles Screen is constructed as follows*:

Ethical Principle 1: Principle of protection of life. The worker's highest obligation is that of protecting human life. This principle takes precedence over all others.

Ethical Principle 2: Principle of equality and inequality. Persons of equal status and power have the right to be treated equally while those of unequal status "have the right to be treated differently if the inequality is relevant to the issue in question" (p. 61). In other words, worker interventions should acknowledge the differential levels of status and power among the people involved. For example, a child does not have the same degree of power in a relationship as does an adult. Therefore, a

---

*Reproduced by permission of publisher, F. E. Peacock Publishers, Inc., Itasca, Illinois. From Frank M. Loewenburg & Ralph Dolgoff, *Ethical Decisions for Social Work Practice,* 1992 copyright, p. 60–62.

worker would need to take greater care to protect the child than to protect the parent's privacy and confidentiality in the case of an abusive relationship.

Ethical Principle 3: Principle of autonomy and freedom. The worker's decisions and interventions should respect the client's right to make his or her own choices and should be directed toward developing the client's ability to function independently.

Ethical Principle 4: Principle of least harm. The worker should intervene in the manner that creates the least harm for the client.

Ethical Principle 5: Principle of quality of life. The worker should make choices that enhance the quality of life for the client and others.

Ethical Principle 6: Principle of privacy and confidentiality. The worker should respect client confidentiality and the privacy of the client's home, personal life, and information.

Ethical Principle 7: Principle of truthfulness and full disclosure. The worker should report all information fully and accurately to clients and others who are authorized to have access to such information (Loewenberg & Dolgoff, 1992, pp. 60–62).

Within this framework, each principle takes precedence over the ones listed below it in the hierarchy. In cases in which various ethical principles are in conflict, this model suggests that the worker should identify the specific ethical principles involved and make decisions in accordance with the issue that is highest in the hierarchy. Read the following case example with the Ethical Principles Screen in mind.

Case Example 4

Sophie is 13-year-old who smokes marijuana several times per day, shoplifts regularly, and engages in unprotected sex with multiple partners. Her parents allow her to come and go as she pleases and on many occasions give her no curfew for coming home at night. It is not clear how much Sophie's parents know about her activities. Sophie says her parents know what she has been doing, but at the same time she does not want you or anyone else to discuss these issues with them. The ethical dilemma for staff members in the agency is whether to inform Sophie's parents of her activities, and, if so, how much information should be shared.

Exercise 7.8

Using the Ethical Principles Screen as your guide, discuss the ethical issues involved in Sophie's situation. Starting at the top of the hierarchy and working your way to the bottom, consider each ethical principle and whether it applies to this situation. After working your way through the Ethical Principles Screen, decide upon and describe a course of action that you believe would be advisable, remembering that those principles at the top of the hierarchy should carry more weight in your decision than should those toward the bottom of the hierarchy.

Exercise 7.9

Now apply the Ethical Principles Screen to a situation drawn from your own experience in your fieldwork. (You might wish to use a situation that you described in Exercise 7.6). Identify the issues involved in the situation as represented in the Ethical Principles Screen, working your way from the top of the hierarchy to the bottom as you did in the previous exercise. Paying attention to the placement of each of the principles relative to one another in the hierarchy, decide upon and describe a course of action that takes into account the priorities of the Ethical Principles Screen.

How effective was the Ethical Principles Screen in helping you to weigh appropriately the various issues involved in this particular situation? What were the strengths and weaknesses of this approach?

Although the Ethical Principles Screen is a very helpful tool in making many ethical decisions, it is not in itself a complete decision-making model. Numerous ethical decision-making models have been suggested within the human services literature (e.g., Corey, Corey, & Callanan, 1998; Gross & Capuzzi, 1999; Kitchener, 1984; Forester-Miller & Davis, 1996). Corey, Corey, and Callanan identify the following steps in ethical decision-making based upon their analysis of several ethical decision-making models**:

1. Identify the problem or dilemma.
2. Identify the potential issues involved.
3. Review the relevant ethical guidelines.
4. Know the applicable laws and regulations.
5. Obtain consultation.
6. Consider possible courses of action.
7. Enumerate consequences of various courses of action.
8. Decide on the best course of action (1998, p. 15–16).

These decision-making steps suggest that careful thinking about ethical issues precedes ethical action. After clarifying the various ethical and legal issues involved, you must imagine alternative scenarios and their probable outcomes before selecting a course of action. Thus, abstract considerations about ethical principles in Steps 1–4 must be seen in the light of pragmatic reality in Steps 6–8. An important component in accomplishing both of these tasks effectively lies in Step 5, seeking consultation.

**From *Issues and Ethics in the Helping Professions,* 5th edition, by G. Corey, M. S. Corey, and P. Callanan 1998. Used with permission of Wadsworth Publishing, a division of International Thomson Publishing. Fax 800-730-2215.

Discussing the situation with an experienced, respected colleague is an invaluable method of clarifying your own thoughts, brainstorming alternatives, and gaining the perspective of a less-involved but knowledgeable person.

Exercise 7.10

Identify a decision involving competing ethical principles and/or values that you have either dealt with directly or at least have been aware of in your field agency. Use the ethical decision-making steps outlined above to think through this situation carefully. Write below your responses to each step of the process.

Step 1: Identify the problem or dilemma.

Step 2: Identify the potential issues involved.

Step 3: Review the relevant ethical guidelines.

Step 4: Know the applicable laws and regulations.

Step 5: Obtain consultation. (If you have discussed the situation with a colleague or colleagues, summarize these discussions. If not, identify one or two individuals whom you might approach to discuss the situation, then continue through Steps 6–8.)

Step 6. Consider possible courses of action.

Step 7. Consider consequences of various decisions.

Step 8. Decide on the best course of action.

As a final step in evaluating your selected course of action, you might refer again to the Ethical Principles Screen. Is your selected course of action consistent with the hierarchy of principles reflected there? If not, you may wish to review the decision-making steps to make sure that you are comfortable with your decision.

A Student's Reflections on Ethical Decision-Making

I am coming to the conclusion that most of the ethical issues in my organization have more to do with acts of omission that acts of commission. What I mean is that it is not as though the staff does bad or harmful things, but at times they just don't do what they can to be there for the clients. The social worker in skilled nursing, for example, spends no time with the residents. I know she is busy with meetings and paperwork and such, but what does the word "social" in social work mean? I feel that she could decide to mark out a couple of hours a week in her schedule just to go around and visit the residents. Until this internship, I had a different, more narrow understanding of ethical issues. Now I see that every minute of the day you are making ethical decisions just by deciding how to spend your time.

CONCLUSION

Learning to identify ethical issues and handle them appropriately is a professional skill that you will continue to develop throughout your career. Such issues can be extremely complex and difficult to sort out. Unfortunately, there is not always one clear "right thing to do." The best guiding principles to take away from this discussion are to (1) maintain constant awareness of the ethical and legal standards as well as the values of the profession as you go about your daily activities, (2) work on cultivating your sensitivity to ethically questionable situations so that you can recognize them early and prevent problems, and (3) once you have identified a difficult issue, seek consultation from experienced professionals as you make decisions regarding how to proceed. As you gain more experience and practice, your ability to weigh competing principles and make ethically responsible decisions will grow.

Your Reflections on Ethical, Legal, and Values Issues

REFERENCES

Alle-Corliss, L., & Alle-Corliss, R. (1998). *Human service agencies: An orientation to fieldwork.* Pacific Grove, CA: Brooks/Cole.

Blair, N. (1996). Law, ethics, and the human services worker. In H. Harris & D. Maloney (Eds.), *Human services: Contemporary issues and trends* (pp. 183–193). Pacific Grove, CA: Brooks/Cole.

Corey, G., Corey, M., & Callanan, P. (1998). *Issues and ethics in the helping professions* (5th ed.). Pacific Grove, CA: Brooks/Cole.

Cournoyer, B. (1996). *The social work skills workbook* (2nd ed.). Pacific Grove, CA: Brooks/Cole.

Egan, G. (1994). *The skilled helper: A problem-management approach to helping.* Pacific Grove, CA: Brooks/Cole.

Ethical Standards of Human Service Professionals (1996). *Human Service Education 16(1),* 11–17.

Everstein, D., & Everstein, L. (1983). *People in crisis: Strategic therapeutic interventions.* New York: Brunner/Mazel.

Forester-Miller, H., & Davis, T. (1996). *A practitioner's guide to ethical decision-making.* Alexandria, VA: American Counseling Association.

Gross, D., & Capuzzi, D. (1999). Ethical and legal issues in counseling and psychotherapy. In D. Gross & D. Capuzzi (Eds.), *Counseling and psychotherapy: Theories and interventions* (2nd ed., pp. 43–64). Englewood Cliffs, NJ: Prentice-Hall.

Kitchener, K. (1984). Initiation, critical evaluation, and ethical principles: The foundation for ethical decisions in counseling psychology. *Counseling Psychologist, 12*(3), 43–55.

Kurzman, P. (1995). Professional liability and malpractice. In R. Edwards & J. Hopps (Eds.), *Encyclopedia of social work* (pp. 1921–1927). Washington, DC: NASW Press.

Loewenberg, F., & Dolgoff, R. (1992). *Ethical decisions for social work practice* (4th ed.). Itasca, IL: F.E. Peacock.

Mehr, J. (1998). *Human services: Concepts and intervention strategies.* Boston: Allyn & Bacon.

Neukrug, E. (1994). *Theory, practice, and trends in human services: An overview of an emerging profession.* Pacific Grove, CA: Brooks/Cole.

Patterson, L., & Eisenberg, S. (1983). *The counseling process* (3rd ed.). Hopewell, NJ: Houghton Mifflin.

Reamer, F. (1995). Ethics and values. In R. Edwards & J. Hopps (Eds.), *Encyclopedia of social work* (pp. 893–902). Washington, DC: NASW Press.

Schram, B., & Mandell, B. (1994). *An introduction to human services: Policy and practice.* Boston: Allyn & Bacon.

Woodside, M., & McClam, T. (1998). *An introduction to human services.* Pacific Grove. CA: Brooks/Cole.

Writing and Reporting Within Your Field Agency

Students choosing human services as a profession generally make this choice based upon their interest in people, human relationships, and social problems. No one enters the field because they like to do paperwork or present information before a group. Nevertheless, both types of reporting are frequently required of students in their internships and throughout their careers. As a student, whether you are in an administrative role or a direct service role, you will find that both written and oral reports of various kinds are a required part of your daily activities. Unlike some aspects of your internship role, this may be a task for which you have little formal preparation. This chapter will focus on the written and oral reporting required in human service agencies.

A Student's Reflections on Written Reporting

I can't say that nobody warned me because in all of my human services classes the importance of writing skills has been preached. Somehow, though, I never realized before how much writing is involved in this type of work. I am spared a lot of it as a student because I can't document things in client records, but my supervisor has to write down every single thing that happens with a case. If she even has a short conversation with a worker from another agency who is working with one of our clients, she has to write it down in the chart. The record includes everything from these little three-sentence "Talked with therapist . . ." type of statements to long, formal court reports that are read by the judge and everyone else working with the case. Although there is a lot of written work, I can also see how many of the staff have built up speed and efficiency in getting it done.

WRITING IN HUMAN SERVICE SETTINGS

Writing is an important component of virtually all human service organizations. Although you have become accustomed to the writing demands of being a student, the internship may be your first exposure to the writing demands of the profession. These demands can be particularly difficult because they often involve new styles and formats of writing and very limited time in which to produce a finished product. Adding to this difficulty is perhaps your preference to spend the limited time you have available in your internship on tasks that you value more, such as spending time with clients or working on projects to which you have been assigned. All of these factors conspire to put writing low on the priority list for many students as well as for many professionals. Like many students, human service workers often consider the paperwork demands to be too time-consuming and sometimes become resentful of these responsibilities as well as resistant to carrying them out. As a result, lack of attention to record-keeping has become a serious problem in many human service settings (Kagle, 1995).

Despite the many factors that might distract you from your writing responsibilities within your internship, do not allow yourself to develop bad habits now. Not only is writing an integral part of virtually any human services position, it is also a skill that you can develop and improve only with practice. Even if your writing is not required by the agency, you would be wise to write mock reports and ask for your supervisor's feedback on them in order to develop your skills. Also, as you go about your work, take the time to read, analyze, and evaluate the written work of others that you might have access to through client records, agency reports, and other documents. Learn what you can by carefully examining those writing samples that strike you as particularly well written and incorporate what you learn into your own writing skills (Truitt, 1991).

Exercise 8.1

During your internship, you have probably had the opportunity to try your hand at paperwork by completing various forms and written reports within the agency. You have probably also had the opportunity to observe staff members' habits regarding their paperwork. What have your reactions been to the paperwork requirements of your agency? How have you handled your responsibilities in this area? What methods do you see staff members using to handle the writing demands of their work?

The following case examples will illustrate just how serious problems with documentation can become. Professionals who do not attend to the paperwork demands of their jobs can unintentionally harm the clients they are attempting to serve, as well as their own careers.

Case Examples: Two Cautionary Tales on the Importance of Documentation

Harold had been employed as a human services worker for over 20 years in a sheltered workshop for adults with developmental disabilities. He was an effective worker and was well liked by his clients and colleagues as well by the agency's administrators. Over the years, he had become increasingly lax in maintaining his paperwork. His documentation of client contacts generally ran weeks, sometimes even months, behind. Harold told himself and others that "this paperwork stuff just isn't important. I'm too busy doing things that really matter to push a pencil for a bunch of bureaucrats." A review of the agency by state authorities found many documentation deficiencies for which they officially cited the agency. As a result of this reprimand, certain components of the agency's funding were even jeopardized. Upon closer examination, agency administrators found that many of the documentation deficiencies were Harold's. When Harold's next performance evaluation was done, his supervisor rated his work unsatisfactory in this area and recommended that he receive no pay increases at all until he improved his work habits. In fact, Harold was even issued a warning that he could be terminated unless he changed his paperwork habits. He was given 90 days to show sufficient progress.

Harold felt angry and demoralized by his evaluation, yet there was little he could say in his own defense because his supervisor's evaluation was based upon objective data regarding his performance. Harold had deluded himself into believing that the paperwork requirements of his job really did not matter. Clearly, they mattered a great deal to the overall well-being of the agency. In some situations, client well-being can also be affected by poor documentation habits, as the following example illustrates.

Myra was a therapist in an adult mental health program that dealt with severely and persistently mentally ill clients. Myra had been working with a client, Carl, attempting to secure a placement for him in a long-term residential treatment program. She had made a formal referral to a particular program, which she documented in Carl's record but did not identify by name.

One Friday while Myra was away on vacation, Carl came into the program in a severe crisis. He needed an immediate residential placement outside his home. Because he was an unscheduled "walk-in," Sheila, the human service intern in the agency, was asked to work with Carl. Sheila's supervisor suggested that the residential treatment program to which he had already been referred might be the best bet for a speedy placement for Carl, especially if they should happen to have an emergency or respite bed. An additional advantage with this plan was that Carl could be spared a potentially upsetting transition from one program to another at some later time. Unfortunately, Carl was so disoriented that he could not remember the name of the program or anything about it. Because the program was not named or clearly identified in the record, there was no way to locate it except through trial and error. Sheila spent much of the afternoon on the telephone calling various programs before identifying the one to which Carl had been referred. By this time, Carl had grown increasingly agitated and upset, but finally Sheila succeeded in making a placement for him. The process, however, took hours longer than it might have had the documentation in the chart been more detailed.

Harold's career was harmed by his lack of attention to his paperwork responsibilities. Carl's psychological state suffered (and probably Sheila's as well!) due to a careless omission in the record. Such events are all too common in human service settings, causing unnecessary hardship, inefficiency, and problems.

WRITTEN REPORTS RELATED TO DIRECT SERVICES TO CLIENTS

Writing about services to clients is probably the most frequent type of writing within most human service agencies. Documenting information about clients and the work that is conducted with them and on their behalf is an essential part of a human service worker's professional role. Knowing the purposes of documentation, general guidelines for doing such writing well, and some of the most common types of writing regarding services to clients will better equip you to meet the challenges of writing during your fieldwork.

Purposes of Documentation

If you see your writing responsibilities as just a useless bureaucratic requirement, you will of course be reluctant to complete your written work. Far from being merely bureaucratic "red tape," written documentation is an important component in the delivery of quality of services to clients and serves to protect your clients, your agency, and yourself. With this understanding, you will probably have greater motivation to do your written reporting and do it well. Following are some of the most important purposes of written reports and records regarding services to clients:

- *Written documents are necessary to provide continuity of services and care.* (Kagle, 1996; Wilson, 1980). Clients sometimes need services when their usual worker is not available. The worker may be sick, attending a professional meeting, on vacation, or may have even left the agency for another position. A clear, thorough record provides the information necessary for a different worker to provide services for the client efficiently and knowledgeably. In residential care facilities, the transition between workers occurs several times a day as workers report for various shifts. Smooth transitions rely upon workers leaving clear, thorough, and accurate information for one another. Another issue in continuity of care occurs when clients seek services in other agencies. With the client's permission, recorded information in such cases can be exchanged between the two agencies in order to reduce duplication of effort.
- *Written records help to ensure that workers have developed systematic plans and interventions for their clients' care* (Wilson, 1980). Writing a client care or treatment plan in which the client's history is summarized, current problems are identified, goals are set, and action plans are developed can be a time-consuming task. Nevertheless, developing such a plan in writing requires the worker to compile and organize relevant data, interpret it, and make decisions about how to proceed. Otherwise, overworked professionals might be tempted to rush from one client to the next without giving sufficient thought to planning services and interventions.
- *Written records provide legal documentation of worker activity* (Kagle, 1996; Wilson, 1980). Human service agencies and workers are held accountable for the services they provide. If a client should challenge the quality of care he or she received, the case record is considered legal evidence of professional activity. By the same token, action that is taken by the worker but not documented may be considered as not having occurred at all. Therefore, it is extremely important to document any tasks that you agreed to do on behalf of a client, all interventions you make on behalf of a client, and any tasks or assignment the client agreed to complete. It is also important to document outcomes of interventions, including progress made as well as less successful efforts.
- *Written records provide legal documentation of client needs* (Kagle, 1996). Just as it is important to document your own activity, it is important to document client be-

haviors, concerns, needs, and requests. Workers and agencies are sometimes put in the difficult position of defending why they did or did not take certain actions on behalf of a client. A record that shows that worker activity was based on observations of client behavior and/or concerns that the client reported offers a clear, credible rationale for the worker's decisions and actions.

■ *Agency funding may be based upon the documented services delivered* (Kagle, 1996; Wilson, 1980). When agencies seek funding, whether it be from private grants or public revenues, they must submit statistics regarding the services delivered within the agency. Any service provided but not documented cannot be legitimately reported. Some funding sources do periodic site visits and conduct record reviews to verify that services which have been reported or billed were actually provided. The axiom, "If it isn't documented, it didn't happen," is applied in such reviews. As pointed out in the case example earlier in the chapter, agency funding can be significantly harmed by a worker's failure to record his or her work.

■ *Written documents are used as a method to monitor the quality of the services delivered within an agency* (Kagle, 1996; Wilson, 1980). Agencies often use peer review committees to examine client records for information about the quality of the agency's work. Records might be reviewed to gather such information as whether services were delivered promptly, the number of worker contacts required to resolve client difficulties, the effectiveness of interventions being used by workers, and so on. Supervisors may also use records in similar ways to monitor the work of the staff members for whom they are responsible.

Exercise 8.2

For which of the preceding purposes does your agency use written documentation?

How are records used in your agency to enhance the quality of service delivery? Do you see ways in which your agency could derive greater benefits from its documentation efforts?

What suggestions might you make for improving the paperwork requirements? What suggestions could you make about how documentation might be more effectively or efficiently done?

Ground Rules for Documentation

Before you begin your work with documentation, it will be helpful to understand some of the ground rules and guiding principles for writing about client care. Some of these ground rules are:

1. For client records to be useful they must be kept up-to-date. Write promptly to document any activity in the case so that another worker can be ready to assist the client in your absence, if necessary.

2. Be concise and to the point but detailed enough to ensure that the necessary information is conveyed. Workers do not have time to read pages and pages of details. A good client record will present the relevant information in a clear, concise, organized fashion so that it can be read and understood fairly quickly. By the same token, the earlier case examples (see "Case Examples: Two Cautionary Tales on the Importance of Documentation") illustrated how important the inclusion of certain key details can be. Such critical details can be included without adding significantly to the length of the report.

3. Take care to document particularly clearly client needs and worker activities in an effort to meet those needs. These are central features in the work of any human service intervention.

4. Although most clients generally do not ask to see their records, clients legally do have access to any information regarding them in most settings. Be mindful of this whenever you write. Be honest and factual but tactful in your writing.

5. Do not use slang terminology in describing clients or their concerns. Although clients might use slang terminology in describing their own concerns, professionals should always use appropriate professional language. For example, in one report a student described a client with a drinking problem as a "lush." Although the client had used this term to describe herself, the student should have used more professional language.

6. Avoid subjective or judgmental language (Wilson, 1980). Use factual description instead. For example, rather than saying, "This client's house was disgustingly filthy," you might say instead, "There were eight large dogs inside the house. Dog feces and fleas were on the floor and furniture throughout the house."

7. Avoid diagnostic labeling (Wilson, 1980). Use factual descriptions instead. For example, instead of saying, "This client is anorexic," it would be better to say, "This 21 year old client is 5'6" tall, weighs 85 lbs. and may have an eating disorder."

8. Use "person first" language and avoid referring to people as a diagnosis. For example, rather than reporting that the "client is a paranoid schizophrenic," it is preferable to say, "the client has paranoid schizophrenia."

9. Avoid or minimize sensationalizing language and phraseology such as "the client suffers from . . .", "he is a victim of . . .", or "she was stricken with . . . ." Use more straightforward, descriptive language such as the client "has" or "has been diagnosed with" a given illness or disorder.

10. Be clear in your writing about the sources of your information. Specify whether information is based on the client's report, your direct observation, or the report of a third party.

11. Clarify factual information as distinct from your impressions, opinions, and interpretations of the facts, although doing so might only involve a change of a few words. For example, it is one thing to say, "It is my impression that this client is not very motivated to find a job," and another to say, "This client is not motivated to find a job."

12. Always pay attention to the basic mechanics of good writing, including spelling, sentence structure, grammar, and organization. Once you have entered information into a client's record, it may be read by many professionals for years to come. Impressions are formed about you, your abilities, and your professionalism based upon the quality of your writing.

A Student's Reflections on Documentation

There was a major issue with a resident's family today. The doctor had changed a woman's medication during the course of the day. At the transition between second- and third-shift staffs, no one told the third-shift staff about the change. They gave the woman her old medicine that she had been taking. She then had a really bad reaction to the mix of medications. Understandably, her son was very upset about this. Although it was written in the record, the staff are supposed to write notes about any changes directly onto the log that stays on the counter in plain view in the nurses' station. It isn't efficient for the staff to sit down with a big stack of charts every shift to see what's been going on. This time the log system didn't work.

Common Types of Written Reports Regarding Direct Client Care

Each human service organization has its own unique methods of recording information about client care. Despite the differences from one organization to the next, there are in general a few common types of reporting that are present in most situations. Each of these types is briefly discussed here.

Client Care Plans (Treatment Plans)

Client care plans are reports that describe client needs and identify goals that the client and worker are striving to attain. The care plan, as the name suggests, lays out a plan for achieving the goals. Responsibilities of both the worker and the client are identified in the plan and tentative dates may be set for completion of various tasks or subgoals (Mehr, 1998). This report might be called a "treatment plan" in mental health facilities, medical facilities, or in other settings that follow a medical model.

The care plan is often one of the most lengthy documents in a client's record. Even so, it is important to remember that being concise and brief is a highly valued trait in all documentation. When a worker is reviewing a record to refresh his or her memory about a case, or when a new worker picks up the record to become acquainted with a client before a first contact, neither has time to read pages and pages of detailed information. Finding the right balance between thoroughness and brevity takes practice and experience. By reading records within your agency, you can get some idea of what the norms are about the length of various documents within that particular setting.

Subheadings are generally used to organize the report and to help readers quickly identify any portion of the report they might be seeking. At minimum, the client care report should include the following information:

- *Identifying information* consists of such information as client name, gender, age, race, family status, employer/school, physical appearance, socioeconomic status, and so on.
- *Presenting problem* records the immediate concern(s) that brought the client to the worker's attention.
- *History of problem* explains how the presenting problem(s) developed over time.
- *Social history* includes information regarding the client's development, family background, employment and/or educational history, legal history, medical history, and services previously received. All of these topics might not be relevant to all clients. Also, depending upon the agency's mission and goals, certain topics would get greater emphasis than others. For example, medical settings would pay more attention to medical history, while corrections facilities would place greater emphasis on legal history.
- *Goals* identify the outcomes the client and worker will pursue in their work together.
- *Plan* specifies the methods, activities, and strategies that will be employed to achieve the goals.

Exercise 8.3

Using the six headings above, write a care plan for a client in your setting. If you are not in a direct client-care position, either interview a worker about a client or read a client's record to gather the information necessary to construct the report.

Identifying Information

Presenting Problem(s)

History of Presenting Problem

Social History

Goals

Client Contact Notes (Progress Notes)

Contact notes are written after each encounter a worker has with a client, including office visits, home visits, school visits, telephone contacts, and so on. In residential care facilities, a note is written about each client at the end of each shift. Client contact notes tend to be fairly brief, perhaps a paragraph or two. They are sometimes referred to as "progress notes" or "summary notes."

Although various formats are used, a particularly useful and common framework for organizing such a note is the SOAP approach. SOAP is an acronym that stands for the following components of a note:

- *Subjective (S)*. This portion of the note documents the client's report or that of the client's family or significant others—that is, the client's or other's subjective experience. For example, "Mrs. S. reported that her husband punched her and slapped her repeatedly last night and that she called the police this morning."
- *Objective (O)*. This portion of the note reports what the worker directly observed in the contact with the client. For example, "The left side of Mrs. S.'s face was swollen and badly bruised at the time of the interview. Using the phone in my office, she phoned the police to ask additional questions regarding how they would handle her report."
- *Assessment (A)*. The assessment gives the worker an opportunity to share any impressions, opinions, or interpretations relevant to the client contact. For example, "Mrs. S.'s face appeared as though it could have some broken bones. It is my impression that she may be in imminent danger if she returns home following her report to the police."
- *Plan (P)*. The worker reports any plans made with the client during the contact and/or interventions the worker plans to make on the client's behalf. The plan should include tasks to be completed by the client as well as by the worker. For example, "Mrs. S. planned to go directly to her doctor's office after our visit. I gave Mrs. S. information about Domestic Violence Services and encouraged her to call them as soon as possible, which she agreed to do. Secured Mrs. S.'s permission to contact the shelter to inquire about openings and admission procedures. Will call Mrs. S. at the home of her friend this afternoon to finalize plans for her immediate safety."

The use of subheadings is not generally required for contact notes unless they are fairly lengthy. Whether or not the subheadings are used, the SOAP format provides an excellent structure to organize the worker's thoughts as well as to provide an organizational structure for the note itself.

Exercise 8.4

Using the SOAP structure, write a summary note of a particular contact that you have had with a client or a contact that you have observed another worker conduct.

Periodic Summaries

Periodic summaries, sometimes referred to as periodic reviews, are required at regular intervals within most settings. These reviews might be required at quarterly intervals (every three months), six-month intervals, and/or annually. These reports require the worker to look at his or her work with a client over a greater period of time. Such reports are valuable because they require the worker to take stock of the progress that has or has not been made and to evaluate the effectiveness of the interventions being used. In the absence of such reviews, workers can get caught up in the day-to-day efforts being made with numerous clients and lose sight of the "big picture."

Periodic reviews typically include the following:

- the time frame covered by the report
- goals accomplished during this time
- interventions the worker made during this time
- new developments or problems that have arisen for the client during this time
- goals that have yet to be accomplished
- plans for continued service, especially citing any revisions of the original care plan

Reviews are generally written in paragraph form without use of subheadings. Complex cases with many challenges and contacts might have fairly lengthy reviews of a page or more. Most reviews tend to be fairly brief, consisting of a few (1–3) paragraphs.

Exercise 8.5

Write a periodic summary of your helping relationship with a particular client over the course of your internship. If you have not had the opportunity to work with a client over time, either read a client's record or interview a worker regarding a client to gather information to construct such a report. For the purposes of this exercise, you may choose the time interval for your review, with a one month review being the minimum review interval.

Closing Summaries

Closing summaries are written when services end for any reason. The closing summary should reflect an overview of the client's contact with the agency. It should include a statement of the presenting problem(s), a description of services provided, and a statement about the degree of progress made toward the various goals. The closing summary should also include information about the reasons why services are being terminated. In addition, any particularly significant events in the course of the agency's relationship with the client should be reported.

In writing this type of summary, keep in mind that if the client returns to the agency for services in the future, the closing summary should be helpful in reorienting the worker as to the client's status and to previous efforts to assist the client. This objective is especially important, of course, if a different worker assumes responsibility for the case upon the client's return. Also, the closing summary can become especially important if another agency requests information about the client following the termination of services. The closing summary is the most logical source of current information in the client's chart in that it offers a comprehensive picture of the agency's work with the client and represents the most recent documentation of the client's situation (Kagle, 1996; Wilson, 1980). As is obvious from this description, the closing summary is fairly likely to be consulted as an important source of information in the future. Therefore, it is especially important that closing summaries be thorough, accurate, concise, and well written. You may have opportunities to write closing summaries as you terminate with clients toward the end of your internship.

A Student's Reflections on Written Reporting

My supervisor has a pretty big stack of records for clients whose cases need to be closed. The stack has sat around her office for a while, and she has asked me to write a closing summary for each case. When she first asked me to do this, I thought it would be impossible. I had never even met these clients, so how could I write the closing summary? Strange as it may sound, I have been able to do this, and she tells me I have done a good job with it. From reading the client's chart, I identify things like the goals they worked on and the progress they made. In some cases, I can also see why they terminated, but in others they just kind of dropped out or disappeared. I'm learning a lot about the work that they do in this organization from reading and working with the records, and I'm practicing my writing skills as well.

Most students find that the type of writing required in client records takes some practice. Continue to practice writing the various types of reports as much as possible. A good practice exercise is observing another worker's contact with a client and writing a summary. Compare your summary with the one written by the staff member. Discuss with the worker any significant differences that you see between the two summaries. Such discussions will help to clarify any changes you might need to make in your methods of documentation. In most cases, documentation done by students must be cosigned by a supervisor (Simon, 1999). This safeguard helps to ensure that you are getting sufficient instruction about the particulars of how documentation is done within your setting.

WRITTEN REPORTS RELATED TO ADMINISTRATIVE SERVICES

Students and professionals in administrative roles may write as much or more than do those in direct service roles, but their writing tends to be of a different type. If you are in an administrative internship placement, the writing requirements that you encounter might be quite varied. Administrators often write for the purpose of internal communication within the agency as well as for communication with external audiences, such as other agencies, funding sources, community organizations, media outlets, and the community at large. A few of the most common types of reports are discussed in the following sections of this chapter.

A Student's Reflections on Administrative Reporting

As one of my projects I was asked to take charge of a major statistical report that they have to do every year. It involves retrieving data from various computer files, organizing it into tables and charts, and writing a narrative report that describes and interprets the information. The report is important to the agency's accreditation because it documents staff training in every department and division. I am pleased to say that I have been doing pretty well on this project and that I have gotten lots of praise and pats on the back for my work. Who says that paperwork is a thankless task?

Reports About Agency Programs

Reports about the agency and its various programs may be prepared by program directors or coordinators and submitted to agency directors and other administrators for the purpose of intra-agency communication. Such reports might include statistical information regarding the clients served as well as descriptive information about the programs, services, and administrative processes within the unit. These reports often identify strengths and weaknesses in the program or unit and make recommendations for improvements and changes. Program evaluations, in which administrators assess the extent to which program goals have been accomplished and the quality of the agency's functioning in the course of meeting these goals, are routine in most organizations. Such reports are usually written annually and form the basis for program planning and implementation for the next year (Weiner, 1990).

Time Interval Reports

Time interval or periodic reports are required within many agencies. These reports are sometimes referred to as "process evaluations" (Lewis, Lewis, & Souflee, 1991) or "interim reports" (Seidl, 1995) because they provide ongoing information about the agency's activities that is used to monitor how effectively the various units within the organization are achieving the agency's goals, month-by-month or quarter-by-quarter. Each unit within the organization might be required to submit such a report to the agency director monthly or quarterly in order to ensure that each unit's work is "staying on track." Such reports may be fairly comprehensive, covering such topics as number and types of clients served, staff-client ratios, specific services rendered, financial expenditures and acquisitions, special program initiatives, and so on for the time frame of the report.

Reports to the Board of Directors

Reports to the board of directors routinely include financial information and progress reports from various units and/or the organization as a whole. Because the board of directors is responsible for the agency's performance, it must monitor activities closely (Gelman, 1995). Written reports are often used for this purpose. At times, additional types of reports are generated exclusively for the purpose of communicating with the board. Examples of such reports might include formal proposals for new programs, for program modifications, or for policy changes. Such reports are submitted to the board of directors as a first step in seeking board input and approval. Reports of various kinds (both written and oral) are made to the board at most regular board meetings. The by-laws of agencies dictate such meetings at set intervals, such as monthly, every two months, or quarterly.

Exercise 8.6

Do some research in your agency to identify the various types of writing that administrators are required to do. Which of the types of writing discussed here are generated within your setting and at what intervals?

What additional types of reports and writing are required?

Compile samples of the various types of administrative reports and writing required within your agency and critique their usefulness. Does the administrative writing and reporting that is done contribute to the agency's effectiveness? If so, how? If not, discuss what changes might be made to use these efforts more effectively.

Funding Requests

Funding requests are written in agencies of all types. Some agencies may seek funds from tax revenues through filing proposals to various elected boards and/or government representatives. Other agencies may seek funds by making requests to such sources as United Way, private donors, or grant sources, both public and private. Private nonprofit programs might even submit funding requests to corporations. Two common types of documents for fund-raising purposes are case statements and grant requests.

The case statement is an important document in fund-raising from private sources. The case statement serves to cultivate interest and support from potential donors. This document typically includes a description of the organization seeking funds, a convincing argument as to why support is warranted and for what purpose, and an explanation of how contributions will be used to address the needs described. This statement is then used in a number of ways. The document may be used as an outline by volunteers as they speak with potential donors, as an initial mailing to prospective donors prior to a person-to-person visit from a campaign volunteer, or as a handout to audiences following fund-raising presentations (Turner, 1995). The case statement is an important fund-raising tool in those organizations that engage in soliciting donations directly from donors.

Exercise 8.7

What efforts, if any, does your agency make to raise funds from the community? What written materials do they produce toward this end? Explore whether your agency uses case statements in its fund-raising efforts. Locate recent examples of a case statement or other written materials used in fund-raising and critique their effectiveness below.

Over the last two decades, grant writing has become increasingly common in human service agencies. Although the various sources of grants might require somewhat different types of proposals, grant requests generally include the following information: a description of the need or problem that the funding will be used to address, evidence to support the existence of the need or problem, a description of the project for which funding is being sought, the specific goals and objectives of the program, the strategies or plan of action that will be used to achieve the project's goals, the methods that will be used to evaluate the program, and a budget or estimation of the program's cost (Lewis, Lewis, & Souflee, 1991).

Inquire about how you can become involved in any fund-raising and grant-writing efforts in the agency so that you can learn as much about the process as possible. Human service students can considerably enhance their skills and employability by developing expertise in these areas.

Exercise 8.8

What efforts does your agency make to secure funding through grants?

From what specific sources does the agency seek or receive grant money?

Read any grant proposals that your agency has submitted in the past and critique them. What do you see as the strengths and weaknesses of these grant proposals?

Is your agency currently developing a grant proposal? If so, what is the grant for and to what sources is it applying?

If your agency does not pursue grant money, explore and discuss the reasons it refrains from grant-related activity.

A Student's Reflections on Grant Writing

I am working on a grant for a new position in the agency. I have researched and located potential funding sources and am currently engaged in the process of writing the grant itself. The agency needs to set up a one-on-one mentoring program in the community to work on drop-out prevention, but no one has the time or money to get one going so the grant I am working on will try to secure funding for this. It's exciting to be working on a real grant, especially since the grant would create a job I would love to have if it is funded. I have an extra incentive thinking that I may be writing up my own first job opportunity. Whether or not this works out, I now know that I can write a real grant. I think employers will be interested in that.

Public Relations

Public relations has become a critical component of administrators' roles in human services organizations. Written materials comprise one of the most essential cornerstones of the public relations efforts within many agencies. Such activities as writing press releases, creating brochures, and producing agency newsletters are now commonplace within human service agencies of all types and sizes. These efforts are important in maintaining community awareness of programs, marketing the services that the agency offers, and managing the public image of the organization (Weiner, 1990).

Compile written materials that are produced by your agency to enhance its public relations efforts. Read and critique these materials as to their effectiveness. What are the strengths of these materials? What are their weaknesses?

If your agency creates no materials of this type, what are your ideas about why this is the case? How might the agency benefit from the creation and distribution of such materials?

If your agency is one that does not currently produce any public relations materials, you might offer to produce some samples as a special project during your internship. You might try your hand at creating a draft for an agency brochure, for example, and share your efforts with your supervisor and other staff members. If your agency produces such materials, inquire about how you might get involved in these efforts. For example, you might write an article for the next edition of the agency's newsletter. Students are often able to provide positive, fresh, and interesting perspectives on the agency in such articles. Such efforts on your part will give you an excellent opportunity to practice and develop new skills while providing a wonderful service for your agency as well.

ORAL REPORTING

In addition to written reports, oral reports are commonly required of students and professionals in human service settings. Professionals in administrative roles frequently present information to staff groups, community groups, potential funding sources, and boards of directors. Professionals in direct service roles speak before their peers and/or supervisors to present case material, to receive consultation, and to conduct education and training for one another. Such presentations are often done in formal, regularly scheduled meetings sometimes referred to as staffings, case conferences, review team, or grand rounds. These presentations are a valuable part of any organization's functioning and serve a number of purposes. They ensure that workers are receiving adequate oversight and the input of other professionals as they conceptualize and intervene in complex human lives. Additionally, such presentations can constitute a form of continuing education for staff as they learn from hearing their peers discuss various issues.

Exercise 8.10

Under what circumstances have you presented information orally to other professionals, community groups, staff members, and so on during your fieldwork?

How comfortable were you in these situations?

Most students and many professionals need to practice their oral presentation skills as much as possible. Can you identify additional opportunities that you might pursue in order to practice your oral presentation skills during your internship?

A Student's Reflections on Oral Reporting

There was a meeting of the Board of Directors last night, and my supervisor asked me to come for a few minutes and tell them about the work I've been doing. I don't like talking in front of groups, but I went. I took an outline of the different things I have been doing and what I wanted to say about each one. This may seem silly but I had even practiced. As I drove along in the car over the past few days I would talk out loud, pretending I was doing my talk. I think things went well. I had the most trouble at the end when they asked me questions. I hadn't really prepared for this, and I heard myself saying "uh" a thousand times as I tried to get my thoughts together. I can do pretty well when I'm rehearsed, but I need to get more practice talking off the cuff.

Students frequently report that presenting information orally is an anxiety-provoking experience for them, and understandably so. Oral presentations can be particularly challenging because both verbal and nonverbal aspects of the communication must be handled effectively. To enhance your presentation skills, it may be helpful to review some basic principles about presenting information effectively.

To understand the skills involved in oral presentation, it is important to remember that face-to-face interactions carry information in three categories: vocal expression, body language, and verbal expression. One researcher has suggested that as much as 93% of the total message is conveyed through the nonverbal avenues of vocal expressions and body language (Mehrabian, 1981). Vocal expression includes such factors as rate of speech, voice pitch, volume, and tone of voice. To evaluate your effectiveness in this area, you might arrange to record yourself in a meeting or presentation. Either audiotape or videotape would be useful to gather information about your vocal expression. In general, you should strive to speak in a strong voice, loud enough to be heard by everyone in the room without their having to strain to hear you. You should speak slowly enough to be understood but not so slowly that others become bored or impatient. Anxiety can sometimes tighten the vocal chords, causing the speaker's voice pitch to become higher or even creating a quiver in the vocal chords. Anxiety can also influence the speaker to rush through the information, speaking far too quickly to be easily understood. If you feel that anxiety interferes with the quality of your presentations, it may be useful to practice some relaxation methods such as taking slow, deep, even breathes, inhaling slowly through the nose, and exhaling slowly through the mouth (Johnson, 1993). This breathing pattern helps to induce a relaxed response and can be practiced fairly inconspicuously anywhere.

Nonverbal communication includes not only vocal expression but body language and facial expressions as well. Although a great deal of attention is given in human service education to nonverbal communication as it relates to interactions with clients, you might have given less thought to its relevance in other areas of professional communication. Many of the same principles and guidelines that apply to making effective presentations apply to interacting with clients effectively. For example, one of the most significant factors in making effective presentations is facing the person or persons whom you are addressing squarely and making eye contact (Johnson, 1993). As you know, this same communication skill is also required in effective interaction with clients. Despite the fact that a student may maintain eye contact perfectly in a helping interview, it is not unusual to see this same student stare directly down at a pad placed in his lap or twirl her hair anxiously in front of her eyes when presenting information before a group or in supervision. In addition to maintaining eye contact, avoid fidgeting and other anxious habits when presenting. Common examples of such behaviors include shaking the foot, swinging the foreleg, rocking or swaying from side to side, playing with a pen or paper clip, doodling on a pad, twisting a ring or earring, and handling the hair. All of these behaviors convey anxiety and distract listeners from hearing what you have to say. Focus on standing or sitting straight, speaking up, and maintaining eye contact.

Finally, verbal expression is obviously important because it carries the intended content of the message. To present a clear verbal message, you will need to plan and organize what you want to say ahead of time. Whether you are presenting a case in staffing or making a presentation before the agency's board of directors, you need to think ahead of time about the major points you wish to make and prepare notes that can act as prompts to keep you focused and organized as well as to jog your memory, if necessary (Halley, Kopp, & Austin, 1998). Under no circumstances should you simply read your remarks from a prepared draft. Although it may be reassuring to write word-for-word what you want to say, this might then tempt you to just read what you have prepared. It is generally more helpful to speak from notes, even if they are quite detailed.

Although a speaker might be very knowledgeable and well prepared for a presentation, the effectiveness of this presentation can be quickly undermined by counterproductive language habits. For example, the credibility of a presentation is weakened if the speaker repeatedly uses qualifiers such as "I think," "sort of," and "I guess." Similarly, speakers whose sentences are sprinkled with such phrases as "uh," "well," "um," like," and "y'know" are not only difficult to listen to but also come across as lacking in confidence and poise (Weaver, 1993). As suggested earlier, audiotaping or videotaping is perhaps the best way of identifying such language patterns in yourself (Truitt, 1991).

Other guidelines regarding the verbal component of presentations closely parallel those discussed earlier for written communication. Just as in your writing, use correct grammar and avoid the use of slang language. Also avoid using subjective or judgmental language, focusing instead upon factual data to make important points. Acknowledge when you are making a generalization or sharing a personal impression and back up your statement with specific facts and examples. Although you might have strong feelings about a given client or situation, your oral presentation will be more effective if you avoid highly emotional displays, adhering instead to factual, descriptive information. Similarly, a speaker presents as more accurate and credible by avoiding extremes of verbal expression. For example, describing a situation in terms such as "always," "never," "all," or "none" should be avoided unless it is *literally* true (Weaver, 1993). Figurative use of such terms can reduce the speaker's effectiveness by creating the impression that he or she is exaggerating or holds rigid or extreme views. Of even more concern is the possibility that a situation or person may be inaccurately and unfairly described through the use of such terminology.

Since oral presentation skills are clearly quite important, it is a good idea to try to get some objective feedback about your own oral presentation skills during the course of your fieldwork. You might arrange to audiotape or videotape yourself when presenting, then review the tapes to evaluate your skills. It is also helpful to ask for feedback and suggestions from your supervisor and other co-workers who have observed you when presenting information within the agency.

Exercise 8.11

Evaluate yourself as an oral presenter in your fieldwork. If you have not had the opportunity to present information within a group, you might focus on occasions in which you have presented information to your supervisor or other workers individually. Discuss the quality of each of the three components of your oral presentation skills: vocal expression, body language, and verbal expression. What are your strengths in each of these three areas? In what specific ways do you need to improve in each area?

CONCLUSION

Many students are surprised to learn how much time and effort human service professionals put into writing and reporting. Although students are generally well prepared to provide client services, they may be less prepared for the reporting functions of human service worker roles. Skills in verbal communication, nonverbal communication, and writing are required to handle these requirements effectively. These skills can be developed through ongoing effort, practice, motivation, and self-evaluation.

REFERENCES

Gelman, S. (1995). Boards of directors. In R. Edwards & J. Hopps (Eds.), *Encyclopedia of social work* (pp. 305–312). Washington, DC: NASW Press.

Halley, A., Kopp, J., & Austin, M. (1998). *Delivering human services: A learning approach.* New York: Longman.

Johnson, D. (1993). *Reaching out* (5th ed.). Boston, MA: Allyn & Bacon.

Kagle, J. (1996). *Social work records* (2nd ed.). Prospect Heights, IL: Waveland Press.

Kagle, J. (1995). Recording. In R. Edwards & J. Hopps (Eds.), *Encyclopedia of social work* (pp. 2027–2032). Washington, DC: NASW Press.

Lewis, E., Lewis, F., & Souflee, F. (1991). *Management of human services programs* (2nd ed.). Pacific Grove, CA: Brooks/Cole.

Mehr, J. (1998). *Human services: Concepts and intervention strategies* (5th ed.). Boston, MA: Allyn & Bacon.

Mehrabian, A. (1981). *Silent messages: Implicit communication of emotions and attitudes* (3rd ed.). Belmont, CA: Wadsworth.

Seidl, F. (1995). Program evaluation. In R. Edwards & J. Hopps (Eds.), *Encyclopedia of social work* (pp. 1927–1932). Washington, DC: NASW Press.

Simon, E. (1999). Field practicum: Standards, criteria, supervision, and evaluation. In H. Harris & D. Maloney (Eds.), *Human services: Contemporary issues and trends* (pp. 79–96). Boston: Allyn & Bacon.

Truitt, M. (1991). *The supervisor's handbook: Techniques for getting results through others.* Shawnee Mission, KS: National Press Publications.

Turner, J. (1995). Fundraising and philanthropy. In R. Edwards & J. Hopps (Eds.), *Encyclopedia of social work* (pp. 1038–1044). Washington, DC: NASW Press.

Weaver, R. (1993). *Understanding interpersonal communication* (6th ed.). New York: HarperCollins.

Weiner, M. (1990). *Human services management: Analysis and applications.* Belmont, CA: Wadsworth.

Wilson, S. (1980). *Recording: Guidelines for social workers.* New York: The Free Press.

Chapter 9

Managing Your Feelings and Your Stress

A human services internship can feel like an emotional roller coaster. The excitement and stress of being in a new and unknown situation with numerous responsibilities can make an internship in any discipline emotionally challenging. Compounding this situation is the intensely personal and interpersonal nature of most human services internships. Working daily with difficult human problems prompts emotional reactions in even the most experienced professionals. Within the span of just a few minutes you might feel both the gratification of knowing that you have helped a client and the frustration and helplessness of feeling that you have little or nothing to offer. You might experience genuine warmth and empathy for one client followed by irritation and frustration toward the next. All of this is within the scope of a normal day's work in human services. The challenge for you as an intern, and throughout your career, is to manage and deal with these emotions in a manner that is both professional and healthy. This chapter will focus upon specific skills and strategies that will help you to manage your feelings and your stress effectively.

A Student's Reflections on Managing Feelings and Stress

I am busy every single minute. This agency is so understaffed it is pathetic. I took on the responsibility because I'm here to learn, and I thought it would be good for me. Now I think I have way overextended myself. I blew up the other day at Sam, one of the guys who works here. I am in charge of planning a big field trip for all of the kids, but I thought it was going to be a team effort. Sam didn't do the things he agreed to do, which then put me in a crisis. I don't have room for three or four more things on my list of things to do. Not to mention that I have been really worried about one of the girls who has been coming to the program. She seems really down, and I think I have smelled beer on her breath a few times lately. She is only 13 years old and a sweet kid. I know there are a lot of alcohol and drug problems in her family. I want to approach her about it but the time has to be right. The last thing she needs is for me to scare her away from the program. We are her main support right now. I am completely stressed-out.

DEVELOPING SELF-AWARENESS

Perhaps the most fundamental skill in managing your emotions is self-awareness. Self-awareness is the best place to start in managing your emotions because it enables you to recognize your feelings as they occur. Noticing and honestly acknowledging your feelings, reactions, thoughts, and behaviors is the first step in confronting and dealing with your emotions in a responsible, professional manner. Although the focus of this chapter is on managing feelings, you will find that monitoring your thoughts and behaviors can be helpful in recognizing your emotions. This is because your thoughts and behaviors are often important sources of information about your feelings. Thoughts and behaviors can sometimes act as "signposts" leading you to recognize emotions that may be present but less obvious, accompanying or underlying your thoughts and behaviors.

Developing and maintaining self-awareness during your internship will require you to cultivate a nondefensive, open psychological posture. This can be difficult when you are in the midst of anxiety-provoking or other emotionally intense experiences. Throughout the day, as you find yourself becoming tense or anxious, practice simple, unobtrusive relaxation methods such as deep breathing and relaxing your shoulder muscles. Practice self-observation throughout the day, trying to stay in touch with your emotional reactions to various events. Do not feel that you should necessarily act on these emotions or even express them. In fact, an important goal of self-awareness is to gain greater control over your feelings so that you will not act on them in a manner that is unprofessional or counterproductive. Work on simply observing your feelings, thoughts, and behaviors in various situations as a method of learning more about yourself and learning to monitor your emotions.

Exercise 9.1

What are the prevalent emotions that you have observed in yourself during the course of your fieldwork? What are some of the most difficult emotions for you to deal with productively? What specific events or circumstances triggered these emotions in you? What strategies have you used to deal with these emotions productively?

DEVELOPING SELF-UNDERSTANDING

Building upon self-awareness, you can work toward enhancing your self-understanding as well. Self-understanding goes beyond simply identifying your feelings and involves recognizing the sources of those feelings. Let's imagine that you have just participated in a meeting in your field agency that you found to be very uncomfortable. By the end of the meeting you were feeling quite distressed, anxious, and somewhat annoyed. Although this self-awareness is an essential first step in managing your emotions, important questions remain to be answered. Where do these feelings come from? What particular events in the meeting might have triggered these emotions? Were other participants in the meeting inappropriate in their behavior or are your feelings the result of a more personal "sore spot" for you? Do your feelings stem directly from events in the meeting or are other issues in your life a factor? Do the feelings suggest that there is an issue or problem that needs to be resolved? If so, is it an issue within yourself? a problem within your internship? an interpersonal conflict? a combination of all of the above?

While self-awareness relies mostly on self-observation, self-understanding relies on deeper reflection. It may take some careful thought and analysis to identify the issues involved in a particular disturbing event. Private time for reflection, introspection, and just plain thinking is essential in order to develop this type of self-understanding. Writing in a journal or processing significant events using the Integrative Processing Model (discussed in Chapter 4) are ideal tools for developing greater self-understanding. Also, talking with your supervisor or another good active listener can help you to develop greater insight into your emotional reactions.

Exercise 9.2

Identify and discuss an instance during your internship in which you were able to move from an awareness of your feelings (self-awareness) to understanding the source of your feelings (self-understanding). What methods have you used to cultivate self-awareness and self-understanding during your internship?

DEVELOPING ASSERTIVENESS

Once you have understood the source of your feelings, assertiveness can be especially useful at times in managing your emotions effectively. In some cases, examining and understanding your emotions will help you to identify problems that need to be solved or issues that need to be addressed in your internship. In these cases, the next step must involve expressing these needs to others so that problem-solving can occur.

Assertiveness is simply the ability to straightforwardly communicate your thoughts and feelings in a manner that respects your own needs as well as the needs of others (Alberti & Emmons, 1995). As such, assertiveness is an essential tool for human services workers to possess as they engage in dealing with emotions and solving problems. Assertiveness involves the use of clear and confident "I-statements" in expressing what you want, need, or feel. Some people have difficulty allowing themselves to be assertive because they misunderstand what assertiveness is, believing that assertiveness requires them to be selfish, demanding, or rude. These behaviors are not assertive behaviors but aggressive behaviors, which will be discussed later in this chapter. Remember that assertive communication recognizes and respects the rights of others. If you find it difficult to communicate assertively, it might be helpful to practice the empathic-assertive model of communication. Within this model, the speaker first expresses the other person's perspective before expressing his or her own perspective (Lange & Jakubowski, 1976). For example, in speaking with your field supervisor, you might say, "I realize that you are concerned that this project might be too much for me, but I feel very confident that I can do it well. I would very much like to take it on." Note that this statement reflects the feelings of both parties in the conversation. Nonetheless, the speaker makes a clear, straightforward request.

Assertive communication includes some important nonverbal components in addition to the verbal components previously discussed. Even the best assertive statement can be rendered ineffective if accompanied by incongruent nonverbal messages. Assertive nonverbal behavior includes maintaining eye contact, standing or sitting with an erect posture, facing directly the person whom you are addressing, and speaking in a firm, clear voice.

Observing good models of assertive communication has proved to be an effective tool in helping less assertive people become more assertive. Simply watching another person handle a situation assertively can be a powerful educational experience, potentially expanding your own behavioral repertoire. Take a moment and try to identify a good role model for assertive communication, perhaps someone whom you have observed communicating assertively in your fieldwork setting. Imagine yourself behaving in a similar manner.

Exercise 9.3

Discuss at least one situation during your internship that you handled assertively. How effective were you in maintaining an assertive stance? How might your assertive response have been improved? What lessons can you learn about assertive communication from the model that you selected in the preceding paragraph?

In contrast to assertiveness, both nonassertive behavior and aggressive behavior are ineffective means of dealing with your emotions and do not lead to problem-solving. Nonassertive or passive people tend not to speak up once they identify important needs and feelings. Instead, they are likely to keep their feelings to themselves or perhaps even hide their feelings from themselves. As a result, the problematic situation continues and possibly even grows worse. In addition to the original problem, the individual tends to develop an extra burden of anger and resentment as the troubling situation continues. Consequently, over time, the feelings of the nonassertive person can grow even more intense and difficult to manage. The nonassertive person who does speak up about a concern may be more likely to talk to a third party rather than speak directly with the person who needs to hear the concern. This, too, tends to create more problems rather than solve them, as the following example illustrates.

Case Example: A Nonassertive Response to a Problem

During her internship, Latoya had been uncomfortable with a particular co-worker, Shirley. Shirley had virtually ignored Latoya throughout her internship. She never once asked Latoya to work with her or even to observe her work. Because Shirley's area of work was of particular interest to Latoya and related to her career goals, this was of real concern to her. After a few weeks, Latoya began to express her frustration and irritation about Shirley's "cold shoulder" to co-workers and to other students in the agency. Within a few days, Shirley heard through the grapevine that Latoya had been talking about her. Shirley (who had no difficulty being assertive) confronted Latoya in an appropriate manner, expressing her concern that Latoya had apparently been speaking of her in a derogatory manner to others in the agency. Latoya quickly apologized to Shirley and left the room, still not having the courage to talk with her directly about the situation. For the remainder of the internship, Latoya was even more uncomfortable in Shirley's presence and tried to avoid her as much as possible. Although Latoya otherwise had a very beneficial internship experience, she left at the end of the term with great regrets about how she handled the situation with Shirley. Worst of all, she never had an opportunity to learn about Shirley's job. In this situation, as in many, nonassertive behavior did not solve the problem and, in fact, made matters even more tense and uncomfortable.

Exercise 9.4

Identify and describe a situation in your internship that you handled in a nonassertive manner. Remember that nonassertive behavior might range from doing and saying nothing to communicating only indirectly about your concerns. (If you cannot identify an instance of nonassertive behavior on your own part, think of one in which you have observed another person in your agency behave in this manner.) What was the outcome? How might this situation have been handled more effectively through using assertive communication?

Aggressive communication, though quite different from nonassertive communication, tends also to make situations worse rather than solve problems. Aggressive behavior involves expressing feelings and needs in a manner that attacks others or is disrespectful of others' needs. This pattern may result in the escalation of conflict rather than in problem-solving. Also, the person who employs aggressive communication risks alienating others and damaging relationships (Alberti & Emmons, 1995).

Let us imagine that in response to Shirley's assertive statement of anger described earlier, Latoya had gone on the attack. Feeling wronged, imagine her responding by saying, "How dare you accuse me of causing a problem between us! You have done nothing but be hostile and unhelpful toward me since the minute I walked in the door!" Notice that this statement is accusatory and attacking. No "I-statements" are used.

Instead, Shirley is angrily blamed for the situation. This is not an assertive response, but an aggressive response. From this example, it is probably obvious how aggressive responses can escalate problems and block potential avenues of problem-solving.

Exercise 9.5

Identify and describe an instance of aggressive communication that has occurred within your field agency. (If you are not aware of any instances of aggressive communication within your field setting, you may select one from a different context). Your example may be drawn from your own behavior or the behavior of others. What was the outcome of the aggressive communication? How might the situation have been more effectively handled through the use of assertive communication?

As you review the three types of communication—assertive, nonassertive, and aggressive—which do you see as most typically descriptive of you during your internship? In your personal life? In what areas of your life would you benefit from becoming more consistently assertive?

DEVELOPING CONFLICT RESOLUTION SKILLS

Although most people do not enjoy conflict, it is an inevitable part of human relationships. All organizations, including human service organizations, experience conflict (Halley et al., 1998). In human service agencies, conflicts can and do occur between workers in the same agency, between workers from different agencies, and even between workers and clients. When conflicts occur, they can become major stressors until they are resolved. If you should experience conflicts during your internship, try to view them as opportunities to develop and practice conflict resolution skills. Do not let the conflicts fester for long, because they will tend to weigh you down emotionally and distract you from other important areas of potential learning in your internship. Learning to resolve interpersonal conflicts can be an important part of reducing your stress and managing your emotions effectively, both in your internship and later in your career. Developing and practicing these skills is particularly worthwhile because such skills are highly valued in employees in virtually any work setting. Moreover, conflict resolution skills are especially valuable to human services workers because they form the basis of many human services strategies and interventions (Hagen, 1999).

A Student's Reflections on Conflict Resolution

In my family conflict is something to be avoided. It's not OK to be openly angry or even to disagree. For many years, I thought that I must have a perfect family because we never had any conflict. Because of my background, now I have a great respect for people who are able to handle conflict well. One thing that has impressed me about this agency is how up front everyone is with their feelings. If they disagree with one another, they say so. Usually it is a friendly disagreement, but at times it can even get emotional. The staff seems committed to talking things through and working things out. Although I haven't personally been involved in a conflict situation here, I am learning a lot from just watching these people. They practice what this profession preaches about the importance of communication, respect, and finding win-win solutions.

Your ability to resolve conflicts will rely in part upon your mastery of skills discussed earlier in this chapter. Being aware of your feelings, understanding yourself, and communicating assertively are foundational skills in resolving conflict. In conflict resolution, you will use these skills to state what you want and how you feel. It is equally important that the other person(s) do the same and that you listen respectfully to their needs, wants, and feelings. Negotiations should focus on efforts to find win-win solutions to the extent possible.

Win-win solutions are those that satisfy the needs of all parties. Win-win solutions require that the parties involved engage in a problem-solving process in which the needs and interests of all parties are explored and possible solutions that might satisfy the interests of all parties are brainstormed and considered. Win-lose negotiations occur when the parties doggedly pursue an outcome more favorable to themselves than to the other persons, employing such strategies as tests of will, manipulation, or power plays. Although win-win conflict resolution takes time and patience, it is most beneficial to preserving goodwill in long-term relationships (Johnson, 1993). Therefore, this method is advisable for co-workers in human services agencies who are likely to work together for some time. Although you may be in the organization a relatively brief time, the relationships you form there can have lasting effects on your career. It is wise to treat them as long-term relationships.

Effective conflict resolution requires effective anger management. Letting anger take control can easily destroy the possibility of constructive problem-solving. Managing your anger does not mean that you do not feel it or express it. Rather it means that you contain it so that your efforts toward problem-solving can stay on task. If you feel the need to express your anger, do so constructively, using assertive rather than aggressive communication (see "Developing Assertiveness" in this chapter). Also, when you feel that you need to express your anger, it is helpful to strive toward expressing positive feelings as well. Doing so helps to balance the possible negative effects that your anger might have on the problem-solving process. Expressing any genuine feelings of appreciation, respect, or liking of the other party can be helpful in keeping the lines of communication open when angry feelings become particularly intense (Johnson, 1993). Managing anger effectively also requires the ability to receive and accept anger from the other party (Weeks, 1992). Listening to the other person's expressions of anger without retaliating with your own anger and defensiveness can be extremely difficult to do but is a critical skill when trying to keep constructive negotiations and problem-solving on track.

In order to resolve conflicts constructively, you need to avoid some of the most common obstacles to conflict resolution. One common obstacle is trying to rush toward a solution too quickly. When the parties hurry to resolve the problem, they tend to take premature "bottom-line" stances or put forth pat proposals for solutions that they then hold onto rigidly. One or all individuals can become fixed on a certain position that they believe "must" happen. Taking this stance before thoroughly exploring and understanding the situation, the interests of the various parties, and the range of options that might be available clearly interferes with optimal conflict resolution. Therefore, as you try to resolve conflicts, try to stay as flexible as possible about specific solutions and encourage the other parties to do likewise.

Similarly, trying to find an easy solution through simply "distributing" benefits or resources can seem to be a straightforward and efficient method of resolving conflict. Such approaches, however, tend to block all parties from exploring how the resources or benefits might be expanded. In other words, if the parties involved simply debate about "how to slice the pie," they may overlook possible strategies that might "make the pie larger."

Finally, trying to apply power to solve the problem is another common, unproductive strategy in conflict resolution. The parties involved may try to settle the dispute by focusing on such issues as which party has more support from administrators, which party would win in a court of law, or who has what "rights" in the situation. Though these points hold some validity, such efforts toward solution create animosity and distract the parties from finding reasonable outcomes in which the interests of all parties might be met (Mayer, 1995).

Exercise 9.6

Think about the methods of dealing with conflict that you have you used during your field placement. (If you have not been involved in a conflict in your fieldwork experience, you may draw upon an example from your personal life for the purposes of this exercise.) Describe the situation, your actions, and the outcome.

Evaluate the effectiveness of your conflict resolution methods in this situation. Based on this critique, in what specific ways might you improve your conflict resolution skills in the future?

Should you experience particularly challenging conflicts during your fieldwork, your field supervisor and/or faculty supervisor should be notified and possibly included in the conflict resolution efforts. Having a third party available as a consultant or possibly as a mediator in the conflict resolution process can be invaluable.

DEVELOPING POSITIVE SELF-TALK

It is difficult being a novice, particularly in a setting in which it might seem that all others are uniformly competent and highly experienced while you are the only "new kid on the block." Sometimes you might feel that no one is ever there to witness your shining moments of success but that your deficiencies are glaringly apparent to all. Comparing yourself to established professionals day after day and repeatedly confronting situations that are new, confusing, and even forbidding at times, can eventually take its toll on your self-esteem. Barker's definition of self-esteem suggests, however, that you can assume direct responsibility for shaping your self-esteem because "an individual's sense of personal worth . . . is derived more from inner thoughts and values than from praise and recognition from others" (1995, p. 340). Long's definition reflects a similar idea as it suggests that self-esteem is based on "*perceived* [emphasis mine] strengths, attributes, and actions" (1996, p. 178). These statements are consistent with cognitive-behavioral theories which suggest that our emotions and behaviors are shaped largely by our thoughts rather than by the situations and events that we experience in and of themselves. In other words, it is not necessarily the events in our lives that upset us emotionally, but what and how we think about those events.

According to these theories, one of the most effective ways to change emotions or behaviors is to change the underlying thought process (Beck, 1976; Ellis, 1962). Consistently thinking thoughts in which you give yourself realistic, accepting, supportive messages is likely to help you feel more confident and relaxed and behave more calmly and appropriately, according to the cognitive-behaviorists. Realistic, accepting, supportive messages are sometimes referred to as "positive self-talk" (Ellis, 1988). Examples of positive self-talk might include such statements as:

- This is challenging but I can handle it.
- My clients are responsible for making changes in their own lives. I cannot do this for them.
- I will do many things right, but at times I will make mistakes. This is normal and acceptable.
- I do not have to be perfect to do a good job.
- I have learned quite a bit about my profession, but I cannot expect to know everything.
- It is OK to say, "I don't know."

Note that positive self-talk does not consist of unrealistic, overblown statements about your abilities or your performance. Likewise, positive self-talk is not simply "positive thinking" in which you insist in your own mind that all will go well (Meier, 1989). These types of statements tend not to be effective because they are unrealistic—that is, not based on reality. Unrealistic messages are neither believable nor reassuring over the long run.

"Negative self-talk" occurs when people give themselves distorted, overly critical, and unrealistic messages about themselves or their situation. Such messages are thought to compound stress reactions and, in some cases, to cause depression and other forms of more serious emotional distress (Ellis, 1988). Common examples of negative self-talk are:

- I must please everyone all of the time.
- If I make a mistake, I am a failure.
- People must give me what I need at all times.
- I should always be available to the people who need me and know what to do for them.
- This is the worst situation imaginable.
- I'll never get it right.
- That client I saw never came back, and it's all my fault.

As these examples illustrate, negative self-talk tends to take some common forms such as "catastrophizing," jumping to illogical conclusions, overgeneralizing, magnifying or minimizing certain circumstances, personalizing, and all-or-nothing thinking (Beck, Rush, Shaw, & Emory, 1979; Freeman, 1990).

Cognitive-behavioral methods have been found to be effective in helping people manage their emotions more effectively (Granvold, 1995). Your work as a human service professional will no doubt present you with many emotionally challenging situations and events. During your internship and throughout your career, you are advised to work on developing positive self-talk and minimizing negative self-talk as an important and effective tool in managing your emotions and your stress in these situations. Your internship gives you an opportunity to observe your mental habits as you deal with various workplace situations and to work on "reprogramming" your self-talk as needed.

Exercise 9.7

Observe your self-talk over the course of a few days in your internship. What messages do you routinely send yourself about yourself, others, and your situation?

As you review these messages, what examples of positive self-talk do you observe? What examples of negative self-talk do you observe?

For each negative self-talk statement, write a more realistic, supportive statement that you can use to replace the negative self-talk.

Changing your self-talk habits will take daily practice and effort. Keeping on your desk or in your journal a list of the positive self-talk statements that you have written in this exercise will serve as a good reminder to practice using them each day.

> ## A Student's Reflections on Self-Talk
>
> If I could start my internship over I would talk more to the other staff members. I eventually did this, but by the time I got warmed up to them there wasn't much time left. As I look back at my first weeks there, I realize I built up a lot of barriers between myself and them in my own mind. I would often think things like, "They don't like me" or "They don't want me around" or "I'm intruding on her turf." Now that I've gotten to know everyone, I see how off target all of this was. I caused myself a lot of stress and missed some opportunities because of my own wrong-headed beliefs about the people I worked with.

DEVELOPING SELF-CONTROL

One of the most challenging aspects of managing your emotions may be learning self-control. Although self-control is not directly discussed a great deal in human service literature, Cournoyer emphasizes its importance, stating , "[Y]ou must be able to manage your emotions, . . . you must manage your opinions, consciously select the words you say, monitor your body movements and gestures, and control your facial expressions and your vocal tone" (1996, p.44–45). All of this is done in service to the principle that it is the client's thoughts, feelings, needs, values, and attitudes, and not our own, that must be primary in the helping relationship. In order to appreciate fully the value of self-control in managing your feelings effectively, it may be useful to think of it in contrast to its direct opposite, impulsivity. Impulsivity may be defined as "the inclination to act suddenly, in response to inner urges, without thought and with little regard to the consequences of the action" (Barker, 1996, p. 181). The human services intern or professional worker who speaks or acts impulsively could potentially do a great deal of damage to others as well as to him or herself.

Maintaining self-control and avoiding impulsivity require a great deal of self-monitoring and hard work. A supervisor early in my career shared an observation with me regarding this that I have often found to be true. After an encounter with a particularly frustrating client, he stated that in human services work you are often tired at the end of the day, not from what you have done or said, but from what you've worked so hard *not* to do or say. Because such self-restraint is often required, it is essential for workers to find acceptable ways to express these unexpressed feelings and frustrations. Talking with your supervisor or other co-workers about your feelings in such situations can provide a much-needed release of frustration and stress.

A Student's Reflections on Self-Control

There is one particular resident at the group home who is always putting his arms around any woman that comes along, trying to hug and kiss them. The staff has been working with him really hard to change this behavior, but anytime a new woman comes into the picture, it seems that he has to learn it all over again. We are all being very consistent with him, telling him to "shake hands to say hello and use words to tell people you like them." I've gotten pretty good at anticipating his moves and avoiding contact, but at times he slips up on me. At these times I feel extremely angry and my first impulse is to yell at him. I keep telling myself that this young man has serious learning problems and that I must be patient with his slow learning process. "Firmness, consistency, kindness, and patience"—I chant this in my head almost like a mantra at times. He will eventually learn, but my patience is wearing thin!

Exercise 9.8

Identify at least one instance in which you needed to exercise self-control in handling your emotions during your internship. Describe the situation and how you handled it. How effective were you in maintaining self-control? What did you learn from this experience that can help you in handling similar situations in the future?

LEARNING TO MANAGE YOUR STRESS

Stress can be experienced as an emotional response and/or a physiological response to a given situation (Robbins, Powers, & Burgess, 1999; Zastrow & Kirst-Ashman, 1997). As an emotional reaction, stress might be experienced mildly as a sense of uneasiness or tension or more severely as a feeling of fear, dread, or anxiety. As a physiological reaction, stress might be experienced through such symptoms as increased heart rate, increased blood pressure, increased perspiration, and/or increased breath rate.

The situations that trigger these reactions are referred to as stressors. Stressors can be external conditions in the environment, such as a conflictual relationship or a pressing deadline for a project. They can also be internal conditions within yourself, such as your attitudes, beliefs, or expectations (Corey & Corey, 1993). As you think about stress in your own life, keep in mind that a certain amount of stress is thought to be healthy and desirable in that it motivates and stimulates a person to act and achieve. A certain amount of stress is an inevitable part of life. Although some stressors might be reduced or eliminated, it is not realistic to expect that stress will be eliminated entirely. Therefore, although it is always useful to examine how certain stressors might be reduced or eliminated, it is equally important to become skillful in managing stress effectively.

Any aspect of life, even the most pleasant, can cause stress from time to time. Consider, for example, that even your most supportive and caring relationships, such as those with family members and friends, probably create some stress for you sometimes.

Certainly some work stress is to be expected as well, especially in a new and demanding situation such as your internship. Although you may be encountering stress in other areas of your life right now, this chapter will focus primarily on understanding and managing work-related stress. The methods of stress management discussed in this chapter, however, can be applied effectively to stress in other areas of your life as well.

Work-related stress is often experienced as an inability to meet the demands of the work environment. You might feel stressed at work from situations such as having too much to do within a limited amount of time, performing a task without sufficient knowledge or preparation, or feeling responsible for an event or situation that seems largely outside your control. Interpersonal relationships at work can also produce stress. Conflict with a co-worker or even constructive criticism from your supervisor might create some stress for you. All of these conditions are considered external stressors, conditions in the environment that trigger a stress response.

While each example noted above involves an environmental stressor, each of these examples might be influenced by an internal stressor as well. You might feel stressed because you expect yourself to get an unrealistic amount of work done within a given time frame or because you expect to know everything about a task before you launch into doing it. These expectations might come from within yourself as much or more than they come from your environment. Similarly, your own beliefs and expectations about relationships might influence your stress level when you are engaged in an interpersonal conflict. For example, you might believe that criticism and interpersonal conflicts are to be avoided and that such events signal some serious shortcoming within yourself. These examples illustrate how stress reactions often involve an intermingling of environmental and attitudinal factors.

Exercise 9.9

What environmental stressors have you experienced during your fieldwork? What internal stressors have you experienced? On the whole, has your stress during the internship more often been caused by external or internal stressors?

Because stress often involves a combination of both external and internal factors, managing stress often calls for a two-pronged approach—dealing with the environment and dealing with your attitudes. A third response is often helpful as well—simply getting your mind off of the stressful situation and engaging in relaxation-inducing activities (Zastrow & Kirst-Ashman, 1997).

In dealing with real stressors in the environment, skills discussed earlier in this chapter can be very useful. Assertive communication and conflict resolution skills can be applied to many situations in such a way that stress can be reduced or even eliminated. For example, communicating directly with your supervisor if you feel that your workload is too heavy or that you are not sufficiently prepared to carry out a particular task opens the door for necessary adjustments to be made. In this case, the stress response has served a useful function in that it has alerted you to a problem that could and should be solved. Therefore, in dealing with your stress, it is always useful to examine whether the stressful situation can be changed or improved. If so, confronting the stressful situation directly is an excellent coping strategy. A part of stress management, however, involves the recognition that all situations in our environments are not amenable to our influence or control. In such situations, it is wise to recognize these limitations and move on to other coping strategies.

Whether or not the situation is changeable, your attitudes often are. When you are feeling stressed, it is always worthwhile to examine whether your own thoughts, beliefs, or expectations might be creating or amplifying a stress response. An excellent strategy for managing such internal stressors is developing positive and realistic self-talk, discussed earlier in this chapter. For example, you might work on becoming more realistic in your expectations about how much work you can complete within a given unit of time. You might acknowledge to yourself that if you wait until you know everything about a given task before you begin it, then you might never do anything in your internship. You might work on reminding yourself that you are in the internship to learn and to be taught, not to prove to others that you already know everything. All of these examples have in common that they do not change the circumstances in which you find yourself, but they do change the manner in which you are perceiving, interpreting, and thinking about these circumstances. Stress, in such cases, might have proved helpful to you in the long run by alerting you to your own unrealistic attitudes that were in need of adjustment.

Exercise 9.10

Specifically, what methods have you used to deal with external stressors? with internal stressors? Cite specific examples of various methods you have used and evaluate their effectiveness.

A final but important strategy for dealing with stress involves directing your attention to other matters. Even when you have made your best efforts at reducing stressors in your environment and forming healthy attitudes, you might find yourself continuing to be preoccupied with and troubled by stress. When you know that you have done all that you can to ameliorate the situation and to modify your own attitudes using the methods described earlier, the time has come to work on changing the focus of your attention.

Sometimes a necessary step in changing your focus is venting your concerns to a friend. All of us have had experiences in which we needed to just talk about a situation and our feelings about it in order to get it off our minds. Talking about your experiences and your feelings with a trusted friend can go a long way toward helping you to move on to other matters. In seminars with fieldwork students, I often examine with them their stress-management strategies. I have found that talking with a friend or co-workers is among the most common stress reliever used by students. A few cautions may be in order, however, in using this strategy. First, although many students use this strategy, they often express concern about and discomfort with the possible confidentiality issues involved. This concern is legitimate and must, of course, be dealt with responsibly in such conversations. Any situations that are in the least questionable in terms of confidentiality should be discussed with your supervisor, faculty liaison, or a co-worker in the agency, rather than with a friend. Many interns also report that they feel most comfortable discussing their work-related stresses with other human service interns who are likely to understand and relate personally to the context of various events. To some extent, field seminars might be used for this purpose.

A second caution about talking out your stress is that talking about the events of the day can simply prolong your attention to these events in a manner that is counterproductive. While it might be important for you to talk about your experiences and feelings, it is equally important to recognize when to stop. Remember that the point in talking about your concern is ultimately to help you shift your focus away from the stressful events. Continued preoccupation with the stressful events through lengthy discussions is likely to only compound your stress response.

You might think of changing your focus away from the stressful events as a method of maintaining a balance between your personal life and your work life. When focusing your attention away from the stressors, the question becomes, "What shall I focus on?" Each person's answer to this question is highly individualized, drawing upon the individual's supports, habits, hobbies, and interests. Exercise, for example, has been found to be a particularly helpful and constructive method of stress reduction. One person, however, might find more relaxation in taking a walk, while another might find playing a vigorous game of tennis to be more relaxing. Experimenting to find the methods that are most effective for you can be a very pleasurable experience in itself. The following is a list of stress-relieving activities that my students have reported to be helpful to them:

- walking
- bike riding
- swimming
- playing sports, such as tennis, racquetball, golf, basketball, and so on
- running
- working out
- reading (not a textbook!)
- spending time with friends or family
- eating dinner out
- praying
- worshipping or attending religious services
- reading scripture
- meditating
- taking a nap
- sitting outdoors
- going to a movie
- lying in the sun
- working puzzles
- listening to music
- playing an instrument
- gardening
- doing yoga
- taking a long bath
- sky diving (!)

This list is shared not only to give you some possible ideas for your own activities but to illustrate the breadth and variety of such stress-relieving possibilities. At times, I have also asked students to identify some less-productive methods of relieving stress that they themselves might have used or observed others use. Watching TV for hours, eating junk food, smoking cigarettes, drinking alcohol, and using drugs are commonly cited poor choices for healthy relaxation. This short list clearly illustrates that all "relaxing" activities are not equally healthy, desirable, or productive in the long run. Obviously, making choices that will lead to long-term health and satisfaction rather than "the quick fix" are advisable.

Exercise 9.11

What stress-relieving activities, if any, are you engaging in during the course of your internship? Which of these activities, if any, might prove to be particularly unhealthy over the long run? Which ones do you see as particularly healthy and productive? What further ideas do you have about methods you might use to manage your stress more productively?

A Student's Reflections on Stress Management

Gradually I have developed a new evening routine since I have been in my internship. It didn't all happen at once but kind of came together a piece at a time. I find that I really need some time to myself in the evenings, whereas evenings used to be primarily a social time for me. I may go out with my friends at some point in the evening, but I need some wind-down time first. My wind-down time is usually just listening to music in my room. Even just 15 minutes of lying on my bed listening to music with my eyes closed helps me to relax and let go of the day. After that I like to take a walk or do something a little bit physical. More than ever before, I am able to enjoy just being alone. I am a very social person so this is kind of strange for me. I think I get "peopled out" at work and need a little break before I can be social again.

CONCLUSION

Learning to manage your feelings and handle stress in ways that are healthy and productive will enhance not only your internship performance but your performance in your career and your general level of satisfaction in life. Your internship gives you an opportunity to develop more effective skills in these areas. By working on greater self-awareness and self-understanding, you can develop more insight into your emotions and behaviors. Assertive communication and conflict resolution skills will be helpful in addressing the various situations that are likely to trigger your more difficult and persistent emotional reactions. Developing positive self-talk and self-control will help you to take responsibility for your own emotional well-being and your own behavior, which can be especially challenging during emotionally intense times. Engaging in relaxation-inducing activities will not only help you to manage your stress but will also add variety and interest to your life. As you learn to manage your day-to-day work stress more effectively, you will be able to enjoy a more balanced life as well as to maintain your energy and enthusiasm for your work.

Your Reflections on Managing Your Feelings and Your Stress

REFERENCES

Alberti, R., & Emmons, M. (1995). *Your perfect right: A guide to assertive living* (7th ed.). San Luis Obispo, CA: Impact.

Barker, R. (1996). *The social work dictionary* (3rd ed.). Silver Spring, MD: National Association of Social Workers.

Beck, A. (1976). *Cognitive therapy and emotional disorders.* New York: International Universities Press.

Beck, A., Rush, A., Shaw, B., & Emery, G. (1979). *Cognitive therapy of depression.* New York: Guilford Press.

Corey, M., & Corey, G. (1998). *Becoming a helper* (3rd. ed.). Pacific Grove, CA: Brooks/Cole.

Cournoyer, B. (1996). *The social work skills workbook* (2nd. ed.). Pacific Grove, CA: Brooks/Cole.

Ellis, A. (1962). *Reason and emotion in psychotherapy.* New York: Lyle Stuart.

Ellis, A. (1988). *How to stubbornly refuse to make yourself miserable about anything—yes, anything!* Secaucus, NJ: Lyle Stuart.

Freeman, A. (1990). Cognitive therapy. In A. Bellack & M. Hersen (Eds.), *Handbook of comparative treatments for adult disorders* (pp. 64–87). New York: John Wiley & Sons.

Granvold, D. (1995). Cognitive treatment. In R. Edwards & J. Hopps (Eds.), *Encyclopedia of social work* (pp. 525–538). Washington, DC: NASW Press.

Hagen, J. (1999). Conflict resolution: An example of using skills in working with groups. In H. Harris & D. Maloney (Eds.), *Human services: Contemporary issues and trends* (2nd ed., pp. 173–179). Boston: Allyn & Bacon.

Halley, A., Kopp, J., & Austin, M. (1998). *Delivering human services: A learning approach.* New York: Longman.

Johnson, D. (1993). *Reaching out: Interpersonal effectiveness and self-actualization* (5th ed.). Boston: Allyn & Bacon.

Lange, A., & Jakubowski, P. (1976). *Responsible assertive behavior:Cognitive/behavioral procedures for trainers.* Champaign, IL: Research Press.

Long, V. (1996). *Facilitating personal growth in self and others.* Pacific Grove, CA: Brooks/Cole.

Mayer, B. (1995). Conflict resolution. In R. Edwards & J. Hopps (Eds.), *Encyclopedia of social work* (pp. 613–622). Washington, DC: NASW Press.

Meier, S. (1989). *The elements of counseling.* Pacific Grove, CA: Brooks/Cole.

Robbins, G., Powers, D., & Burgess, S. (1999). *A wellness way of life.* New York: McGraw-Hill.

Weeks, D. (1992). *The eight essential steps to conflict resolution.* New York: Putnam.

Zastrow, C., & Kirst-Ashman, K. (1997). *Understanding human behavior and the social environment* (4th ed.). Chicago: Nelson-Hall.

Chapter 10

Trouble-Shooting

All students want to be successful in their fieldwork, and by far most are successful. Nevertheless, some students do experience some particularly challenging situations along the way. This is not surprising given the high expectations that an internship places upon students. Consider that, as an intern, you must make sound practice decisions, meaningfully apply classroom learning, demonstrate excellent work habits, produce professional quality writing, relate professionally with a variety of staff members and clients, observe and follow professional ethics, and meet the academic expectations of your human service program. Even the most well-intentioned, highly motivated, and best prepared student can hit stumbling blocks on this challenging path.

This chapter is devoted to informing you about some of the potential stumbling blocks in human services internships and to engage you in thinking about how to prevent them and how to deal with them should they occur. A risk in discussing such problems is that of raising undue anxiety in students. In order to avoid this, there are a few important points that you should keep in mind. Remember that some problems and mistakes are normal parts of life and certainly of student internships. If you are expecting your performance to be perfect, then you are expecting far too much of yourself. Most internships proceed quite smoothly and the problems that students encounter are generally minor. Finally, as you read this chapter, keep in mind the adage, "Forewarned is forearmed." In keeping with this adage, the discussion in this chapter should help you to feel more confident in your ability to avoid any serious problems as well as to resolve those problems that might occur.

A Student's Reflections on Making Mistakes

I have never worried too much about making mistakes because I always learned that the only people who don't make mistakes are the people who don't try to do anything. On the other hand, I have never worked in a human services agency. The stakes are a lot higher when you are working with people instead of objects or products. I do believe I have good training and education. I am not coming in cold. I know what is expected of me in most circumstances. If I remember the ethics and values of human services professionals, try to learn as much as possible, listen to my supervisor, and use common sense, I think I will be OK.

MAKING MISTAKES AND LEARNING FROM THEM

It is not uncommon for students to worry about making mistakes during their internship. Let's take some of the pressure off by acknowledging that almost anyone starting a new job or field placement is going to make a few mistakes. Most students (and workers) learn quickly from their mistakes and do not repeat them. Most mistakes are generally not serious and are easily reparable.

The first step in learning from a mistake is admitting that you have made one. Sometimes it might not be clear to you whether a particular action was a mistake or not. If you have a gnawing, uncomfortable feeling about something you have done or not done, the best policy is to discuss the issue with your supervisor as soon as possible. In this conversation you can get some direct feedback about the matter and consider whether some course of corrective action is necessary.

At times, however, others may perceive that you have made a mistake when you had no idea that there was a problem. In this case, your supervisor or another worker will point the situation out to you. In such a situation, you might be caught off guard and feel defensive. Despite this natural response, try to hear the feedback that is being offered and learn from it. Often, all that is needed to settle the waters is a simple and sincere statement such as, "I'm sorry about the mistake. I would like to correct it if I can." Of course, the statement needs to be followed up quickly with whatever corrective action is appropriate.

Case Example 1

Miguel was assigned to a high school guidance office for his internship. Within the first few weeks of his placement, a 14-year-old male student, Antonio, came into the guidance office several times to speak with someone about a conflict with another student. Miguel's supervisor assigned him to talk with Antonio, and the two of them formed a close relationship. In the fourth week of the internship, Antonio walked into the guidance office with Armando, the student he had been complaining about in previous weeks. The two were in a heated conflict. Miguel met with the two together and tried to mediate their conflict.

When the two boys left the office, nothing had been accomplished and Miguel felt angry and frustrated with Armando. As he thought about the conversation, he began to realize that he had spent most of the time listening to Antonio's point of view and at times pressed Armando to change his behavior. He had spent little time hearing Armando's point of view or examining how Antonio's behavior might have been playing into the problem. Miguel realized he had been ineffective in his work with the two boys and was concerned that he might have alienated Armando from ever entering the counseling office again. He thought he should do something about the situation, but what?

Exercise 10.1

Do you see Miguel's mistake as a reparable mistake? If you were in Miguel's position, what corrective action might you take at this point? What lessons do you think he can learn from this mistake?

MAKING MORE SERIOUS MISTAKES

Although most mistakes are reparable and educational, occasionally there are mistakes of such seriousness that the student may not be given a second chance. In fact, in some situations, students are terminated from their internship for making a very serious mistake or for an ongoing pattern of unprofessional behavior. Being an intern is like being an employee in the agency. Therefore, any behavior that is serious enough to cause an employee to be terminated can result in termination of an intern as well. In addition to the agency's policies regarding termination, academic human services departments that sponsor internships also have policies about intern behaviors that warrant termination from the internship. These policies are generally published, and students are well-informed about them prior to beginning the internship. According to Wilson, behaviors that are commonly cited as resulting in termination of an intern include physically harming a client, inappropriate behavior that is potentially damaging to "the reputation or functioning of the agency and/or its clients," behavior which "constitutes a danger to the student or others," and "illegal or immoral behavior" (1981, p. 198). Agency policies regarding termination of employees commonly cite such behaviors as unexcused absences and/or tardiness, breaking the law, possessing or using drugs on the job, leaving work without the supervisor's permission, violating safety regulations or policies, violating dress code policies, and insubordination (Truitt, 1991). In some situations, the agency and field supervisor may be somewhat more patient with an intern than they would be with an employee because they realize from the outset that the intern is not highly experienced. Nevertheless, more serious violations can result in immediate termination. Less serious violations are usually dealt with through issuing a warning and allowing an opportunity for performance improvement. Interns should be aware, however, that there are behaviors that can seriously jeopardize their internship.

Although discussion of intern termination may raise anxiety for you as a student, it is helpful to discuss this topic in order to ensure that you understand the standards and expectations surrounding this issue. Remember that by far most interns excel in their work and are highly valued by their supervisors and other staff members in the agency. Unfortunately, however, some students can at times be naive about the expectations of the internship and not understand how serious the consequences of certain behaviors can be. When serious problems do occur and a student is terminated, it is an extremely difficult and painful situation for all concerned—the student, the supervisor, and the faculty liaison. Prevention of such a scenario is strongly preferred by all parties.

Case Example 2

Bradley was delighted to be placed in the Student Services Office at a local community college for his internship. His field supervisor was the director of student activities. Bradley was responsible for planning and organizing special events. On one particular occasion, Bradley, under the direction of his supervisor, planned a trip to a nearby town for a concert. Students responded well to the idea, and 25 students signed up for the event within just a few days. On the day of the concert, Bradley, a student worker, and the 25 students loaded into vans and attended the concert. As the group drove back, they stopped at a restaurant near the campus for a late dinner. Bradley was of legal drinking age, and he decided to have a beer with his meal. Several students in the group, including some who were under the legal drinking age, also ordered beer and were served.

The policy of the Student Services Office and of the college was that no alcohol was allowed at functions sponsored by Student Services. Also Student Services employees were expected to respect and enforce the state laws regarding the legal drinking age and to encourage students to respect and obey these laws as well. The philosophy of the Student Services Office also strongly emphasized a clear commitment to providing forms of entertainment and recreation that did not involve alcohol. In addition, Bradley and all of the other human services interns had signed a contract with the academic department that confirmed an understanding and agreement that neither alcohol nor drugs were to be consumed during the internship.

As Bradley and the group enjoyed their meal, a Student Services employee happened to walk into the restaurant. Of course, she saw Bradley drinking with the group of students and informed his field supervisor the next morning of what she had observed. After discussing the situation with Bradley and establishing the facts as previously described, the field supervisor and faculty liaison terminated Bradley from the internship.

Exercise 10.2

Imagine that you are Bradley's supervisor or faculty liaison explaining to him the grounds for his termination. What specific grounds for termination would you cite? What risks, if any, might the Student Services Office and the Human Services Department have incurred if they had allowed Bradley to continue in his internship following the incident described? What lessons can he learn from this situation?

The difficulties that most students encounter in the internship are generally much less serious than Bradley's situation. Nevertheless, even far less serious violations of supervisory expectations can harm how a student is viewed by the supervisor and can indicate significant changes that the student needs to make before assuming the full responsibilities of a professional position.

MAINTAINING GOOD WORK HABITS

Ironically, after years of preparation for entering the professional world, the issues that most often compromise a student's performance in the internship are generally related to simple, basic work habits, not the finer points of academic preparation or professionalism. Although a student might have excellent helping skills and impressive theoretical knowledge, these will likely go unnoticed and unappreciated if the intern is not considered a good worker. In the working world, there are some generally recognized common-sense traits of a good worker in virtually any environment. These traits include:

- arriving at work on time
- staying at work for the full day
- coming to work early and/or staying late occasionally, if this is needed
- attending work reliably
- using good grammar
- dressing appropriately for the position

- using time productively
- being pleasant, energetic, and alert
- relating cordially to co-workers
- writing well
- being flexible
- carrying out responsibilities reliably, promptly, and efficiently

Supervisors value these behaviors in their employees in any type of organization, and human services organizations are no exception. Although you are probably not being paid for your internship, you should treat it like a very desirable job. If you are about to complete your academic program, your supervisor will be evaluating your performance in relation to your readiness to handle an entry-level position in the organization. It should be your goal to demonstrate that you are fully prepared to assume such a position and handle it well. If you enter your internship consistently demonstrating the behaviors described earlier, you are well on your way to at least a satisfactory performance, possibly even an outstanding one.

Case Example 3

Tamara, a senior human services major, performed her internship in a vocational rehabilitation program. Her field supervisor noticed that Tamara arrived 15–20 minutes late the first three days of her internship. Having allowed for an adjustment period, the supervisor mentioned the tardiness on the third day and explained to Tamara that it was important for her to arrive on time, especially because this was a behavior much emphasized with the clients in the program. Tamara apologized and explained that tardiness had always been a problem for her and that it was in no way a reflection of her interest in or commitment to the internship. There was little to no change, however, in Tamara's arrival time in the days following this conversation.

A week later, as the problem continued, the field supervisor brought up the problem in a meeting that they had with Tamara's faculty liaison from her college. The supervisor explained in more detail the various reasons why promptness was important in the organization. In this same meeting, Tamara's faculty liaison also emphasized that promptness and reliability were important expectations of the department and stated her expectation that Tamara would improve her performance in this area during her internship. Nevertheless, the tardiness continued even after this lengthy discussion of the problem.

Tamara at times did excellent work within the agency. She worked well with clients and completed several administrative projects. She had used her computer skills to create a highly effective brochure for the agency and produced a statistical report that was long overdue within the agency. She was praised for her good work and enjoyed a pleasant relationship with her supervisor. Tamara felt that she was working hard in her placement and that she was doing an excellent job.

In her evaluation conference, her supervisor praised her for the strengths she had demonstrated and the work she had performed for the organization. She evaluated her skills highly in certain areas. Tamara was crushed, however, when her supervisor responded with a "No" to the question, "If a position were available within your agency for which this student qualifies, would you recommend her as a candidate for the job?" Her supervisor explained that due to Tamara's habitual tardiness, she would not hire her and could not recommend her for a job within any agency. Tamara was disappointed in her evaluation and phoned her faculty liaison to complain about her "unfair evaluation." She stated that there were employees who came to work late and that she was being singled out unfairly for this offense.

Exercise 10.3

What are your thoughts about Tamara's evaluation and her reaction to it? If you were the faculty liaison responding to Tamara's concern about her evaluation, what might you say?

MAINTAINING APPROPRIATE PROFESSIONAL BOUNDARIES

A frequently cited challenge for human services interns is learning to maintain appropriate professional boundaries (Alle-Corliss & Alle-Corliss, 1998; Royse et al., 1999; Thomlison et al., 1996). Professional boundaries have to do with making sound decisions about what behaviors are and are not prudent within your professional role with clients and colleagues. Boundary issues are involved in questions such as:

- "How much should I tell clients about my own life?"
- "Is it all right to socialize with clients?"
- "How involved should I get in the lives of my clients?"
- "Is it appropriate to accept a date with an employee in my field agency?"
- "Is it appropriate to develop a personal friendship with my supervisor?"

There is evidence that suggests that most human service interns need more training and guidance about such professional boundary issues than they generally receive in their educational programs (Slimp & Burian, 1994).

A common boundary issue that human services interns struggle with is the urge to "rescue" clients. Because human services workers want to help others, it is easy at times to get overinvolved in clients' lives in an effort to rescue them from the distress that they are experiencing (Alle-Corliss & Alle-Corliss, 1998). Efforts to rescue might range from offering false reassurance in an effort to decrease a client's pain at the moment to more extreme urges to take clients home with you or to fight their battles for them in their conflicts with others. Whether mild or extreme, efforts to rescue clients are almost always counterproductive because they rob clients of the opportunity to grow and learn through grappling with their own problems. Clients instead tend either to become dependent upon the worker to handle their problems in the future or to become angry at the worker for not rescuing them successfully. In either case, both the client and the worker suffer in the long run.

In relation to questions about professional boundaries, it is generally best to "err on the side of caution." Some case examples will illustrate and help you to understand professional boundary issues.

Case Example 4

Larry worked in a group home for teenagers with emotional and behavioral problems. One evening while he was sitting in his apartment, the doorbell rang and he found two of the group home residents standing at his door. One of the boys, Carlos, had asked him earlier in the week where he lived. Although he initially hesitated, Larry had told Carlos the name of his apartment complex. When Carlos had asked which apartment he lived in, Larry was afraid it would seem rude not to tell him, so he had quickly blurted out the apartment number, thinking, "Oh, what can it hurt?" As Larry stood at the door looking at the two youths, he remembered and deeply regretted that conversation, but he let them in.

The boys informed him that they had run away from the group home and did not intend to go back. They begged him not to call the group home staff and related stories about how harshly they were treated there. They stated repeatedly that Larry was "the only one we can trust, the only one we can talk to." Larry spent hours with the boys trying to convince them to go back to the group home voluntarily. As time passed, he grew increasingly uncomfortable with his involvement in the situation. Convinced at last that the boys could not be persuaded to return on their own, Larry finally called the group home to inform them of the boys' whereabouts. Unfortunately, 6 hours had elapsed since the boys had left the group home. Staff were upset and worried, and the police had been notified. Larry faced possible disciplinary action within his field agency, possibly even termination.

Exercise 10.4

At what points did Larry have the opportunity to handle this situation more productively?

What professional boundaries issues were involved in this situation and how should Larry have handled them?

What are your thoughts about what might have motivated Larry to handle the situation in the manner that he did?

What lessons can Larry learn from this situation?

Boundary issues can also become involved in interns' relationships with agency staff. Interns have at times been asked out by fellow co-workers, asked to buy expensive products that staff members were selling, and have even been employed to work in businesses owned by their field supervisor during the time frame of their internship. Although such situations usually do not carry the risk of potential harm to clients, they can become quite troublesome for the student, potentially damaging the quality of the student's educational experience as well as the student's relationships with the supervisor and/or co-workers. Should a staff member approach you with any request that seems questionable, inform him or her that you would like some time to think about it before responding, then discuss the situation with your supervisor and/or faculty liaison.

Case Example 5

In the first week of her internship, Tracy learned that a co-worker in her agency was running for the local school board. The co-worker, Mary, asked Tracy if she would be willing to help with her campaign. Tracy liked Mary and had difficulty saying "No" to others, so she agreed to help. Never having been involved in politics, she rationalized that working on a campaign would be educational for her. Because the campaign work was to be done outside the internship and outside the business hours of the agency, it did not occur to Tracy that it could cause problems in her internship.

After two weeks of helping with the campaign and learning more about Mary's positions on issues, Tracy was not sure that she really felt comfortable supporting her as a candidate. She tried to pull out of her commitment subtly and gracefully by explaining that she needed to spend more time on her school assignments. She noticed a cooling of Mary's attitude toward her but decided that this was to be expected.

Tracy began to notice that Mary, as well as other workers, were not asking for help with their work or assigning her tasks as they did in the first weeks of her internship. She felt mildly uncomfortable about the situation but was not particularly worried or concerned about it until a supervisory conference in which her supervisor addressed the issue. Her supervisor said, "I'm concerned that you might be developing a reputation among the staff for being unreliable. I suggested to Mary that she get your help on a project she is doing, and she said that you couldn't be counted on to follow through. Do you have any idea about where this impression might have come from?"

Exercise 10.5

What steps can Tracy take to correct the situation in which she finds herself? What steps might she have taken earlier to avoid the situation? What lessons do you think Tracy needs to learn from this situation?

GUARDING AGAINST DUAL RELATIONSHIPS

The case of Tracy involves a dual relationship with a co-worker. Dual relationships constitute a particular type of boundary issue that is a much-discussed and controversial issue in human services today. Due to the frequency and complexity of this particular problem, it is worthy of special attention here. Dual relationships occur when a helping professional assumes a second role with a client or colleague in addition to the professional role (Barker, 1996). When such relationships occur with clients, they are clearly unethical according to the ethical guidelines of the National Organization for Human Service Education (see the Appendix). Given the clarity of this point, it would seem that professionals should not be confused as to whether it is appropriate to develop a personal friendship, social relationship, or business association with a client. By the same token, it is considered inappropriate for professionals to provide professional services for friends, family members, and other individuals with whom they have a preexisting relationship. Despite the clarity of these guidelines, the real world can have a way of blurring this boundary. For example, if you are only superficially acquainted with someone, is it acceptable to work with them professionally? If a client brings you a small birthday gift, does this mark the beginning of a social relationship? If you know a close family member of the client personally, but not the client himself, does this create a boundary violation? Due to the changing nature of relationships and circumstances over time and "gray areas" like those in the previous examples, workers must be constantly vigilant about the nature of their roles with the clients they serve.

Case Example 6

Mingyu worked as a gymnastics coach at the local YMCA while attending college. There he worked with Don, a talented 12-year-old who was on the YMCA's gymnastics team. While performing his internship as a Juvenile Court counselor, Mingyu was shocked to see Don in juvenile court due to charges of vandalism and theft. Although the situation made Mingyu somewhat uncomfortable, he did not feel there was any real conflict because Don was assigned to a different court counselor. Within a few days, a number of issues began to complicate the picture. For example, in staffing later that week, information was shared about Don's family that Mingyu felt sure Don's family would not want him to know. Also, certain information was shared about the family that Mingyu knew for a fact was false. He suspected that either Don, his parents, or both had lied to his court counselor. As the staffing came to a close, Mingyu's heart sank further when the decision was made to refer Don to a group that Mingyu and his supervisor were co-leading. Recognizing the many dual role issues involved, Mingyu scheduled a meeting with his supervisor immediately to discuss the situation.

Exercise 10.6

What do you see as Mingyu's options in resolving this problem? If you were in his position, how would you resolve the problem?

How should Mingyu handle the conflict between what he knows about Don's family and the information reported by the court counselor?

What role, if any, should he have in any discussions about Don that might occur in the agency?

Fortunately, Mingyu was wise enough to recognize the difficult issues raised by the situation in which he found himself. Talking with his supervisor was exactly the right thing to do. Even his supervisor, however, felt that he needed some guidance in the process of sorting the situation out, prompting a wider discussion of the issue among the staff as a whole. In the long run, Mingyu's raising his concerns brought the entire agency to a higher level of awareness about the potential for dual relationship issues within their work.

Although the ethical problems in dual relationships were first discussed in human service literature in relation to worker-client relationships, the discussion has in recent years expanded to include concerns about other types of dual roles during internships, particularly between interns and supervisors (Bernard & Goodyear, 1992; Keith-Spiegel & Koocher, 1985; Slimp & Burian, 1994). When supervisors and students develop roles and relationships with one another in addition to their supervisory relationship, complications can occur that compromise student learning and in some cases might even result in exploitation of the student.

Some common secondary roles that have been discussed in the human services literature are sexual relationships, social relationships, and business relationships (Slimp & Burian, 1994). Kaiser shares an example of a student whose learning was harmed by a dual social role with her supervisor. Several years after her field experience, the student says,

> [W]e had some difficulties in confronting each other about things. I think maybe we had a secret pact to not to talk about some things. There were times when we did talk about uncomfortable issues. I believe we only went so far, though. I think our social relationship got in the way (Kaiser, 1997, p. 66).

In this particular case, the student was not seriously harmed by the dual relationship nor was she particularly concerned about it at the time. In retrospect, however, she realized that her friendship with her supervisor had created a less-than-ideal learning environment for her. As Kaiser (1997) points out, a dual relationship with a supervisor is problematic because it creates a competing agenda in the supervisory relationship. Once an additional agenda is introduced, there is the risk that the goals of that agenda become primary, while the goals of supervision and student learning become secondary.

Social relationships between supervisors and students can cause harm, but the problems created by the development of sexual relationships, therapeutic relationships, and business relationships are generally seen as even more potentially damaging. Kitchener (1988) suggests that dual relationships are more problematic when (1) the two relationships have very different expectations, (2) the obligations of the two relationships are very divergent, and/or (3) there is a wide differential in power and prestige between the two individuals. While a social relationship may be somewhat similar to the supervisory relationship, sexual relations, psychotherapy, and business dealings have markedly different expectations and obligations. Therefore, according to Kitchener's framework, one would expect the risks of harm in such dual relationships to be greater. Also, since there is a considerable power differential between students and their supervisors, Kitchener's model would suggest that there is, in all cases of dual relationships between students and supervisors, a significant risk of students being harmed. The following case example illustrates how a dual relationship involving a business deal between an intern and her supervisor harmed the quality of the intern's experience.

Case Example 7

Elena had been searching for a much-needed larger house for months prior to beginning her internship at the local homeless shelter. After her internship began, Elena spoke with co-workers often about various houses she had seen with her realtor. Disappointment and frustration mounted as the weeks rolled by and a satisfactory house was yet to be found.

Meanwhile, Elena's internship was going well. She was learning a tremendous amount from her supervisor, Grace, an older woman with extensive experience whom Elena admired enormously. On two occasions she had visited Grace's home, once for a staff party and once to drop off a report. On both occasions Elena had admired Grace's home and had even commented, "If only we could find a place like this." Therefore, Elena was delighted when Grace announced that she and her husband planned to sell their house and move into a townhouse. Elena immediately and enthusiastically indicated her interest in the house. Grace had mixed feelings about the situation. She knew that there could be trouble in doing business with her student, but it pleased her to think about Elena and her family enjoying the house. Also, she knew what a difficult time Elena had been having in finding a suitable place. Within days, Elena contacted Grace's realtor and viewed the house with her family. Shortly thereafter, she made an offer on the house. The offer was a good one financially, and Grace, with some misgivings, decided to accept it.

Over the next few weeks, Elena and Grace increasingly spent supervision time, as well as any other free moments at work, talking about the house and the details of the sale. Each "supervisory session" ended with the mutual promise that next time they would stay on task, but each week there seemed to be more pressing issues to deal with about the house. Matters grew worse when conflicts arose between them about the conditions of the sale. Elena began to request repairs that Grace felt were unreasonable. Elena felt put out by Grace's asking for an extension on the closing date because construction on the townhouse that she and her husband were building was running behind. Tension and mutual mistrust began to develop, and both Grace and Elena began to avoid each other as much as possible. By the end of the placement, Grace and Elena were spending limited time with one another, and Elena was informally seeking supervision on a "catch-as-catch-can" basis from other staff members.

Exercise 10. 7

At what point in the supervisory relationship did a clear dual role develop between Elena and Grace? What responsibility did each individual bear in ensuring that the integrity of the supervisory relationship was preserved? How might Grace and Elena have handled the situation differently in order to avoid the dual relationship?

MAINTAINING A BALANCE BETWEEN YOUR WORK LIFE AND YOUR PERSONAL LIFE

The students in Case Examples 4 to 7 had difficulty maintaining appropriate boundaries in their professional relationships. Another type of boundary issue that interns sometimes struggle with is the boundary between their work lives and their personal lives.

In dealing with the demands of the internship, students can find it difficult to maintain the right balance between their work responsibilities and their personal lives. The typical day of the college student usually includes a few classes with some breaks during the day for running errands, doing chores, spending time with friends or family, studying, or just simply resting. The internship, like a job, does not offer this flexibility. Students must figure out how to reorganize their priorities and their time to allow for what is usually at least an 8-hour-a-day job. Because the internship can at times be stressful and demanding, it is also a particularly important time for students to take care of themselves, physically and emotionally. Finding the right balance between the demands of a career and the demands of one's personal life involves a lifelong process of learning and self-discipline for many human services professionals. The internship is an excellent time to begin cultivating good habits in managing these boundaries.

A Student's Reflections on Managing Boundaries Between Work and Personal Life

There are things that I have practically considered it my God-given right to do every day that I have just had to give up. Talking on the phone to friends, watching TV, and even reading the newspaper have become luxuries occasionally indulged in, not everyday occurrences. Although doing this internship has been a real stretch for my schedule, it has forced me to sort out what my priorities are. My evenings I devote pretty much to family until the children are in bed. I do manage to do a little reading and such occasionally while they are still up, especially if they have lots of homework, too. My husband has taken over some of the errands and such that I used to do during the day. Occasionally, I might try to do a couple of things during lunch, but otherwise I keep my day exclusively focused on my internship. My goal after I graduate is to get a full-time job, so I see this as way to start making that transition. But I won't pretend that it's easy!

Although most students strike a healthy balance between their internship and their personal lives, some are swept into a pattern of working far too much. These students might consistently stay at the office late, arrive early, and take work home with them. Even when they are "off," they may still be working mentally, thinking about problems at work or planning the next day in their minds. Because such hard-working, highly dedicated behavior is often praised and rewarded in our society, the student who succumbs to this pattern can easily feel that this is the right thing to do. Through their own work habits, supervisors and co-workers might even covertly and unintentionally communicate that this level of time commitment is expected from others.

Although you might initially feel good about yourself for working so hard, this working style can potentially lead to negative consequences over the long run. Human services workers must practice good "self care" in order to maintain their energy. Personal time is necessary to take care of yourself physically, emotionally, socially, and spiritually. Even so, taking time out for even the most basic essentials of self-care, such as sufficient sleep and nutritious meals, is too often overlooked by busy interns. Such self-care habits are absolutely necessary, however, because without them you will find it difficult to maintain the energy required for a demanding career (Littleton, 1996).

Students who begin their careers working excessively may quickly begin to feel the effects of burnout (Russo, 1993). Burnout occurs among many human services professionals who do not learn to manage the demands of their profession effectively (Cherniss, 1980; Farber, 1983; Freudenberger, 1983). Burnout has been defined as "the process of becoming so emotionally or physically overburdened that one cannot function adequately in a social or work role" (Schram & Mandell, 1994, p. 618). The symptoms of burnout among human services workers include physical and emotional exhaustion, irritability, negativity, decreased productivity, and feelings of detachment and callousness (Sweitzer, 1999). At times, burnout can even lead to more serious symptoms of depression and anxiety and physical symptoms such as headaches and stomach upsets (Schram & Mandell, 1994). Beyond the personal level, agencies' performance and effectiveness can also be adversely affected by burnout within the staff, thus harming the quality of services to clients (Alle-Corliss & Alle-Corliss, 1998).

Obviously, the consequences of burnout are potentially serious. Therefore, you should strive to develop habits during your internship that will help you to avoid burnout. This is important not only to maintain the quality of your life, professionally and personally, but also to assure the quality of services to clients.

Case Example 8

Van was fortunate to work during his internship under the supervision of a professional whom he genuinely admired. Mr. Stevens, Van's supervisor, was the director of a crisis shelter for runaway teens in a major metropolitan area. Mr. Stevens had started the shelter himself and had nurtured it from a struggling, poorly funded, poorly staffed program to a stable, professional, and highly respected organization. Moreover, he had stayed directly involved with the young people who were the clients of the shelter, rather than moving into doing purely administrative work as the agency had grown. Mr. Stevens truly cared about the kids in the program and routinely worked 12–16 hour days. He spent hours each day listening to the teens, counseling them, and helping them make important decisions about their lives. In addition, he was responsible for and effectively carried out a myriad of administrative responsibilities.

In the beginning days of Van's internship, he wanted to be at the shelter every minute. He felt that if he left, he was going to miss something. There was so much he had to learn, from Mr. Stevens as well as from the teens, that he wanted to be there and use every moment wisely. By the end of the first week of his internship, Van had worked 60 hours. He worked a similar schedule during the second and third weeks. By the end of the fourth week of the internship, he was beginning to feel pretty tired. He knew that he needed to establish a different schedule for himself but found it a difficult topic to bring up with Mr. Stevens, especially since Mr. Stevens seemed to be immune to fatigue himself.

During the fourth week of the internship, Van's faculty liaison pointed out to him that he had turned in two assignments late and still had two assignments that he had not turned in at all. Van became very angry and defensive. He argued that the human services department expected far too much of its interns and that there was no way he could live up to everyone's expectation of him. He expressed resentment toward having "all these stupid assignments" and added that he had not even had a conversation with his own roommates or been out with his girlfriend in over two weeks. In closing he stated, "I'm beginning to think I'd be better off as a shoe salesman or something. You don't have to feel guilty at 5:00 if all you're leaving behind is a bunch of shoes."

Exercise 10.8

What incentives are there for Van to change his work habits? What forces are operating to keep him "stuck" in his current behavior patterns? What aspects of Van's situation can you personally identify with? How have you dealt with such boundary issues in your own internship?

In order to avoid overworking and burnout, try to find ways that you can take care of yourself daily. If you are struggling with this issue, review the section entitled "Learning to Manage Your Stress" in Chapter 9. A number of ideas are discussed there that you might find helpful.

Exercise 10.9

To what extent are you maintaining a healthy balance between your internship and your personal life? In what ways, if any, has this been a challenge for you? In what ways do you need to improve your self-care in order to be at your best for your internship?

A Student's Reflections on Stress and Personal Boundaries

Yesterday I had my first exposure to intake. I helped do intakes all afternoon. It was basically 4 hours of dealing with the most depressing human pain, misery, and poverty that I have ever encountered. It was tough to enjoy my nice comfortable home and cozy gas heat when I got home knowing about all those people who can't pay their gas bills and don't have anything. I thought about it all evening and even dreamed about it when I finally went to sleep. I confessed this rather casually to my supervisor this morning. Upon hearing about my night, she shut her office door, sat me down, put her hands on my shoulders and said firmly but quietly and kindly, "You can't take this home with you. If we all did that, we would go insane and we wouldn't be any help to anybody." I appreciated her way of putting this. To help others I must let go of the misery, including my own guilt, and take care of myself. Looking at it in this way will help me exercise the discipline I need in order to avoid getting too involved in my work.

In contrast to those students who have trouble keeping their internship within its proper boundaries, other students might have the opposite problem. These students have difficulty keeping their personal lives within proper boundaries. This might be the case because they are completing their internship at a particularly difficult time in their personal lives. If so, these students might come to the internship worried about their personal problems. Very practical aspects of some types of problems can also directly interfere with the internship's demands. For example, even ordinary tasks, such as arriving on time, might be a challenge for the student who has unreliable child care or a sick relative to care for. Students in extremely demanding personal situations during their internships may be well advised to seek some counseling during this time. If you should find that your personal situation is too demanding, it may be necessary to withdraw from the internship. In most situations, however, this extreme is not warranted. Throughout a human service worker's career, conflicts between personal issues and career have to be managed. Eventually, all responsible workers must learn to do this effectively. Although it may not be ideal to have to begin learning this difficult lesson during your internship, it is virtually inevitable that you will confront that challenge sooner or later. You might look upon the internship as an opportunity to start this learning.

Many students struggle with keeping their personal lives contained during the internship, not because of any particular personal problems they might be having, but due to limited coping skills in dealing with everyday concerns. These students might benefit from such relatively straightforward measures as improving their organizational skills, learning time-management skills, or developing greater assertiveness in their personal relationships.

Case Example 9

Lea was the kind of person whose friends always came to her for help. In fact, it was this observation about herself that had led her into considering a human services career. At times, she wondered if she attracted "needy people." It seemed there was always someone who needed a ride someplace, a friend who was in a crisis and needed to talk, or a family conflict that she needed to mediate. Although Lea had never been the strongest student in the classroom, she knew that she would excel in her internship. She was a helper by nature and was willing to do anything to help anyone. She was looking forward to her internship because she knew she would succeed.

Two weeks into her internship, after she had fallen asleep in a staff meeting, her supervisor pulled her aside to talk. She pointed out that Lea had come in late on two mornings due to oversleeping, had left early on Friday to take a friend to the airport, and had often seemed to have low energy during the work day. Among her supervisor's other comments, she said, "It's beginning to seem that this internship isn't very important to you."

Lea protested that this was not the case. She explained that her roommate was having some problems with her boyfriend, and that consequently, she and her roommate were up talking until the wee hours of the morning almost every night. She explained that she simply had not been getting enough sleep and that she was indeed very serious about her internship.

Lea's supervisor communicated clearly to her that her performance needed to improve in the areas they had discussed. Lea left the conversation disappointed in herself but also feeling somewhat angry and misunderstood. After all, she was just being a good friend to her roommate and had been about the business of helping someone who needed it. Isn't that what human services is all about? Why, she wondered, couldn't her supervisor just understand that and give her a break?

Exercise 10.10

Based on the information in the case example, what understanding does Lea seem to have about what "helping" is?

In order to change her own behavior, Lea might need to rethink her ideas about helping. How might the concept of personal boundaries be useful to her in this regard?

What skills and/or behaviors does she need to apply in her personal relationships in order to establish better boundaries between her professional life and her personal life?

Much of what beginning human service professionals learn in an internship has to do with boundary issues. The internship marks the beginning of what is often a long-term challenge to find the proper balance and boundaries between your personal life and your professional life. Finding this balance is an important element in learning to take care of yourself as a professional. Another element in learning to take care of yourself as a professional involves maintaining physical safety while on the job.

KEEPING SAFE

Unfortunately, acts of violence have increased in society in recent years. Consistent with this disturbing trend, incidents involving physical aggression against human service workers have also increased (Newhill, 1992, 1995). Research suggests that certain populations that human service workers serve tend to be associated with a higher probability of aggressive behavior than are others. One of the best predictors of violent behavior is previous episodes of violent behavior (Star, 1984). Newhill, for example, found that "being young and male, having a history of substance abuse and weapons possession or a criminal record, and having a history of violent behavior [were] associated empirically with a higher probability of being violent in the future" (1995, p. 633). Also, certain symptoms of mental illness, such as paranoid delusions and command hallucinations (i.e., hallucinations in which the individual is ordered to perform certain acts), have a higher association with violent behavior (Newhill, 1992).

Beyond certain client features, workplace conditions can also create increased vulnerability to violence. Budget cuts and decreased staffing in certain environments, for example, have been found to increase the risk of violent incidents (Hiratsuka, 1988; Schulz, 1987, 1989; Petrie, Lawson, & Hollender, 1982). Schultz (1987, 1989) found that violence against workers was common within certain types of settings, such as corrections, health and mental health, and services for the disabled.

Such information suggests that you should be vigilant in gathering information from your supervisor about potential risks that might exist in your particular setting as well as information about how to minimize and deal with these risks. Knowing as much as possible about the populations that your agency serves, as well as its security and safety measures, will help you to take whatever preventive measures are appropriate and prudent within your setting. Ask your supervisor about any aggressive incidents that might have occurred in the past. Also inquire about any staff training or operating policies that are in place to protect workers. In some settings in which there is a particularly high risk of clients becoming aggressive or out of control, workers are trained in specialized methods, such as use of physical restraints. Also, some agencies flag certain cases as "high risk" by placing a color sticker or other marker on the client's record. This serves to alert workers to use special precautions when working with this client (Griffin, 1995).

Although all agencies may not provide formal training for workers on dealing with aggressive clients or flag high-risk cases, many do routinely advise workers on certain practices that can minimize their vulnerability. Workers may be advised, for example, to arrange their offices such that they are seated between the client and the door. Often, agencies advise workers to perform certain types of potentially high-risk tasks, such as home visits, in pairs. Some agencies even encourage workers to leave their office doors slightly ajar when meeting with clients in the agency. This enables nearby workers to be at least somewhat aware of activities within the office and makes it easier for the worker who is in the office to summon help or leave the room.

As a student, you should be reassured by the fact that supervisors usually avoid intentionally involving students in high-risk situations. In some settings, however, a certain amount of risk is everpresent and unavoidable. Also, the unexpected can occur in any setting. Therefore, it is wise to prepare yourself for the possibility of dealing with an aggressive client, even if such an event seems unlikely.

If, despite everyone's best efforts at prevention, you should find yourself in the presence of a client whose behavior seems threatening, certain behaviors on your part can help to de-escalate the situation. Try to stay calm, speaking slowly in a normal conversational tone. Be respectful toward the client and listen empathetically to his or her feelings. Avoid arguing with the client or responding angrily (Royse et al., 1996). When responding to the client verbally, keep in mind that either passivity or coerciveness on your part may result in the client becoming more upset. Assertive communication can be effective in setting appropriate limits and providing a much-needed sense of structure and safety for a client who is feeling out of control. Even so, take care not to come across as controlling or demanding as this will likely intensify the client's anger and anxiety (Murdach, 1993). If you feel that the situation is imminently dangerous, make an excuse to leave the room. Offer to get a drink for the client or say that you need to get something for yourself (Royse et al., 1996). Once you have left the room, alert other workers immediately so that the client is not left unsupervised (Faiver et al., 1995).

Nonverbal communication is, of course, at least as important as verbal communication in dealing with aggressive clients. Do not touch the client and avoid making too much eye contact. Give the client more than the usual amount of personal space, if possible. Avoid making sudden movements and explain whatever movements you need to make. Making an explanatory statement, such as, "I'm going to pick up this folder now so I can look at my notes" can help prevent an agitated client from becoming even more alarmed, suspicious, or upset (Shea, 1988). Sit down while speaking with the client and encourage the client to sit down as well. Standing is generally interpreted as a more aggressive physical stance (Royse et al., 1996).

Finally, use common sense and trust your instincts in such situations. When going into situations that seem risky, have someone come along with you. This is a particularly good idea when making home visits or venturing into high-crime areas. Also, keep in mind that in particularly high-risk situations, law enforcement officers can be asked to accompany workers. Do not hesitate to ask for support if you are feeling uneasy about a situation. Workers who have been involved in violent episodes frequently report after the event that they had a "gut-level feeling" that the situation could be dangerous. In such situations, even if there has been no real or direct threat, it is important to tell someone and take precautionary measures (Safety First, 1998). Unfortunately, rather than asking for help in such situations, workers too often proceed alone, feeling too embarrassed to ask for assistance or even to admit that they are fearful (Newhill, 1995). As with some other topics that have been discussed in this book, it is better to "err on the side of caution." Asking for support when you need it is not a sign of weakness or failure but a sign of professional wisdom.

Exercise 10.11

What is your typical pattern of responding to people who are aggressive or threatening? Based on the information discussed in this chapter, how might you need to modify your typical response in order to intervene effectively with aggressive clients?

What risk factors for violent behavior, if any, exist within your agency and/or the populations it serves? What measures does your agency use to help ensure worker safety?

A Student's Reflections on Personal Safety

A child became very angry today and slammed a door really hard. A staff member was standing just inside the door with his hand inside the doorframe. It was obvious that the girl was trying to slam his hand in the door because we saw her glance at his hand right before she slammed the door. It was a heavy steel door and could have done some serious damage, especially with the force that was behind it when she slammed it shut. Fortunately, the staff member was really alert and quickly pulled his hand back. It was almost like he was reading her mind when she glanced at his hand in the door. If he had not been so alert and so attuned to her thinking, he could be missing some fingers right now or at least have some broken bones.

CONCLUSION

Even the most ideal students in the most ideal settings can have some difficulties occasionally in their internships. This chapter emphasizes that awareness of potential trouble spots can enable you to act wisely in avoiding them.

You can avoid many problems by observing good basic work habits, by maintaining healthy boundaries within your relationships, by striking a responsible balance between the demands of your work and those of your personal life, and by doing all that you can to keep yourself safe.

Your Reflections on Trouble-Shooting

REFERENCES

Alle-Corliss, L., & Alle-Corliss, R. (1998). *Human service agencies: An orientation to fieldwork.* Pacific Grove, CA: Brooks/Cole.

Bernard, J., & Goodyear, R. (1992). *Fundamentals of clinical supervision.* Boston: Allyn & Bacon.

Barker, R. (1996). *The social work dictionary.* Washington, DC: NASW Press.

Cherniss, C. (1980). *Staff burnout: Job stress in the human services.* Beverly Hills, CA: Sage Publications.

Farber, B. (1983). Introduction: A critical perspective on burnout. In B. Farber (Ed.), *Stress and burnout in the human service professions* (pp. 1–22). New York: Pergamon Press.

Faiver, C., Eisengart, S., & Colonna, R. (1995). *The counselor intern's handbook.* Pacific Grove, CA: Brooks/Cole.

Freudenberger, H. (1983). Burnout: Contemporary issues, trends and concerns. In B. Farber (Ed.), *Stress and burnout in the human service professions* (pp. 23–28). New York: Pergamon Press.

Griffin, W. (1995). Social worker and agency safety. In R. Edwards & J. Hopps (Eds.), *Encyclopedia of social work* (pp. 2293–2305). Washington, DC: NASW Press.

Hiratsuka, J. (1988, September). Attacks by clients threaten social workers. *NASW News,* p. 3.

Kaiser, T. (1997). *Supervisory relationships: Exploring the human element.* Pacific Grove, CA: Brooks/Cole.

Keith-Spiegel, P., & Koocher, G. (1985). *Ethics in psychology.* Hillsdale, NJ: Random House.

Kitchener, K. (1988). Dual role relationships: What makes them so problematic? *Journal of Counseling and Development, 67,* 217–221.

Littleton, N. (1996). Personal qualities in a successful human services career. In H. Harris & D. Maloney (Eds.), *Human services: Contemporary issues and trends.* Boston, MA: Allyn & Bacon.

Murdach, A. (1993). Working with potentially assaultive clients. *Health and Social Work, 18,* 307–312.

Newhill, C. (1992). Assessing danger to others in clinical social work practice. *Social Service Review, 66,* 64–84.

Newhill, C. (1995). Client violence toward social workers: A practice and policy concern for the 1990s. *Social Work, 40,* 631–636.

Petrie, W., Lawson, E., & Hollender, M. (1982). Violence in geriatric patients. *Journal of the American Medical Association, 248,* 443–444.

Royse, D., Dhooper, S., & Rompf, E. (1996). *Field instruction: A guide for social work students* (2nd. ed.). White Plains, NY: Longman.

Royse, D., Dhooper, S., & Rompf, E. (1999). *Field instruction: A guide for social work students* (3rd. ed.). White Plains, NY: Longman.

Russo, J. (1993). *Serving and surviving as a human-service worker.* Prospect Heights, IL: Waveland Press.

Safety first is primary rule on job. *NASW News,* (November, 1998), 4.

Schram, B., & Mandell, B. (1994). *An introduction to human services: Policy and practice.* New York: Macmillan

Schulz, L. (1987). The social worker as a victim of violence. *Social Casework, 68,* 240–244.

Schulz, L. (1989). The victimization of social workers. *Journal of Independent Social Work, 3,* 51–63.

Shea, S. (1988). *Psychiatric interviewing: The art of understanding.* Philadelphia, PA: W.B. Saunders.

Slimp, P., & Burian, B. (1994). Multiple role relationships during internship: Consequences and recommendations. *Professional Psychology: Research and Practice, 25,* 39–45.

Star, B. (1984). Patient violence/ therapist safety. *Social work, 29,* 225–230.

Sweitzer, H. (1999). Burnout. In H. Harris & D. Maloney (Eds.), *Human services: Contemporary issues and trends* (2nd ed., pp. 209–225). Boston: Allyn & Bacon.

Thomlison, B., Rogers, G., Collins, D., & Grinnell, R. (1996). *The social work practicum: An access guide* (2nd ed.). Itasca, IL: F. E. Peacock Publishers.

Truitt, M. (1991). *The supervisor's handbook: Techniques for getting results through others.* Shawnee Mission, KS: National Press Publications.

Wilson, S. (1981). *Field instruction: Techniques for supervisors.* New York: The Free Press.

Chapter 11

Evaluating Your Performance

Near the end of your internship, you and your field supervisor will engage in the important process of formally evaluating your work. Although you have been evaluated throughout your fieldwork, this evaluation has been for the most part "formative." This means that the evaluative feedback that you received was for the purpose of educating you and improving your performance (Bogo & Vayda, 1995). The formal evaluation near or at the end of your internship is, in contrast, a "summative" evaluation. Although summative evaluations are also intended to be educational and can improve your performance, this is not their most immediate objective. Summative evaluations are primarily for the purpose of assessing overall outcomes and performance quality. Such evaluations look at the experience as a whole, identifying strengths and weaknesses (Alle-Corliss & Alle-Corliss, 1998). This chapter will help you understand and engage productively in the summative evaluation process that will occur toward the end of your fieldwork.

A Student's Reflections on Evaluation

My supervisor had me fill out the evaluation on myself, and we compared it to his. We were basically similar on most things, but where we were different, I had evaluated myself lower than he had. He has given me feedback along the way, so the evaluation really wasn't too much new information for me. Overall, what I got out of the evaluation was a good feeling about myself and my work. He was very positive about most things, but I also know the things I need to continue working on. I am pleased to say that I can look at my performance with pride.

UNDERSTANDING THE FINAL EVALUATION

Evaluation has occurred throughout your internship as part of your ongoing supervision and learning process. Therefore, there are usually few surprises for students at the time of final evaluations. This is not always the case, however, due to a number of factors. The thorough and comprehensive nature of the final evaluation process brings the entire internship experience under review. In doing so, patterns, themes, and issues might emerge that were less obvious in the day-to-day work with your supervisor. Also, you might find that although you and your supervisor have discussed your strengths and weaknesses openly over the term, the evaluation calls upon both of you to weigh and balance these areas in relation to one another. In this process, you might find that what one party perceives as a minor issue, the other party may see as quite a significant one, and vice versa. Students and supervisors sometimes find that they weigh the importance of various factors differently and perceive the balance between strengths and weaknesses differently.

An important issue in understanding the evaluation process is accountability. Accountability has to do with the responsibilities of the various parties involved in the evaluation and might be defined as the "state of being answerable to the community, to consumers . . . , or to supervisory groups . . . " or as the "obligation of a profession . . . to provide assurances that its practitioners meet specific standards of competence" (Barker, 1996, p. 3). Kaiser defines accountability as "the process of taking responsibility for one's behavior and for the impact of that behavior on self and others" (1997, p. 14). She further clarifies that accountability involves "a commitment to tell the truth" and "a commitment to take responsible action" (Kaiser, 1997, p. 14). In recent years, concerns about issues of accountability in human service agencies, as well as in educational institutions, have increased. Human service educators and field supervisors bear the responsibility for ensuring that those individuals entering the human service field meet certain minimum standards of competence (Simon, 1999). Therefore, student evaluations, as well as evaluations of practitioners in the field, must be objective and rigorously adhere to certain standards. All parties involved in your final evaluation—you, your supervisor, and your faculty liaison—must understand these issues of accountability in order to ensure that your evaluation is as objective and as fair as possible.

Evaluation is almost always an anxiety-provoking experience, regardless of the quality of your work (Bogo & Vayda, 1995; Wilson, 1981). It might be helpful to recognize that this type of evaluation is not unique to your being a student. All human service workers are evaluated regularly by their supervisors, usually on an annual basis at minimum. Therefore, developing the ability to engage productively in the evaluation process should be seen as part of building your repertoire of professional skills. As you approach your evaluation, your expectations about it and reactions to it will tend to be shaped by your previous experiences with evaluation. The following exercise will help you to identify the expectations that you bring into the evaluation process.

Exercise 11.1

Think of a few instances in the past in which your work has been evaluated. You might draw upon experiences within jobs you have held or within previous fieldwork. To what extent did you experience these evaluations as fair and objective? What are your expectations of your evaluation in this field experience based upon your previous experiences with being evaluated?

As you anticipate your evaluation in this field experience, what aspects of the evaluation process do you feel most anxious about? most confident about?

As a developing professional, you are expected to acquire the ability to evaluate your own performance objectively and participate in your evaluation. In previous evaluations, how actively did you participate in the evaluation process? How comfortable are you with your ability to participate actively in this evaluation?

PREPARING FOR YOUR EVALUATION

As evaluation time approaches, you should prepare yourself to be an active participant in your evaluation by reflecting upon the quality of your performance (Thomlison et al., 1996). The first step of this process will be assembling the various tools that will be used in the assessment process. These tools include the following:

- the learning plan or contract that you developed at the beginning of your internship
- the course syllabus or other document that lists the course objectives
- the student evaluation instrument used by your human service program
- any notes, documentation, tapes, or other concrete materials that reflect your work during the internship
- your calendar or day planner for the time span of your internship

Your learning contract, the course objectives, and the student evaluation instrument function as a collective set of criteria for evaluating your performance (Simon, 1999). To evaluate your performance, focus on each goal or outcome identified in these documents. Try to evaluate your own behavior in relation to each criterion. Referring to your notes, tapes, calendar, and other materials will help you to respond objectively to each item, identifying supporting information for whatever rating you might assign yourself. Without such objective consideration of your own performance and behavior, it is easy to lapse into some common rating errors. A rating error is "the tendency to make a global judgment about a worker's performance and then to perceive all aspects of the worker's performance as consistent with that general judgment" (Kadushin, 1976, p. 287). One of the most common rating errors is the "halo effect." An example of the halo effect occurs when students feel that they have done a good job in their field placement and therefore assume that each and every item on their evaluation will be rated positively, considering any negative feedback as unfair. Although a student's global self-assessment as "good" might be accurate, it becomes an obstacle if this generalization interferes with a closer, more objective look at the relative strengths and weaknesses of the student's performance on particular skills. Looking at objective data regarding your work during the internship can help reduce the tendency toward rating errors and enable you to see your own performance more clearly.

Exercise 11.2

Review the learning contract that you developed for your internship in Chapter 2, "Getting Started." Read each goal, the strategies for accomplishing each goal, and the method of measurement that you identified. Rate your performance on each goal and write a brief supporting statement for each rating, citing specific facts and information to substantiate your rating. In rating yourself, use the following scale:

5. I far surpassed the level of achievement identified in this goal.
4. I slightly surpassed the level of achievement identified in this goal.
3. I satisfactorily achieved this goal.
2. I did not achieve this goal but made some progress toward achieving it.
1. I accomplished little or nothing toward accomplishing this goal.

Exercise 11.3

Review the course goals and objectives for the field experience. These objectives were developed by the faculty of your academic department and reflect the department's expectations of what students should accomplish, achieve, learn, and/or demonstrate during the course of the field experience. Using the rating scale outlined in Exercise 11.2, rate yourself on each course objective and write a brief, factually based statement supporting each rating.

Exercise 11.4

Review the student evaluation instrument that your field supervisor will use to evaluate your performance. Rate your own performance on each item using the same rating scale that your field supervisor will use. (The rating scale is usually included on the instrument itself.) For each item, write a few specific facts to support the rating that you assign to your performance.

Exercise 11.5

Write an open-ended evaluation of your performance in your internship. Structure your writing by organizing your thoughts into the following three paragraphs or topics: (1) your greatest strengths in this field placement, (2) your greatest weaknesses during this field placement, and (3) your goals for professional growth and development when you enter your next field placement or job.

When you have completed Exercises 11.1 to 11.5 thoroughly and thoughtfully, you should be well prepared for your field supervisor's evaluation of you and ready to discuss your evaluation in an objective manner. The final steps in preparing for your final evaluation are (1) considering the roles, perspectives, and obligations of your field supervisor in the final evaluation process, (2) developing an understanding of your evaluation conference with your supervisor, and (3) considering the roles, perspectives, and obligations of your faculty liaison in participating in your evaluation and assigning your final grade.

THE FIELD SUPERVISOR'S PERSPECTIVE

Like students, supervisors also often find evaluation to be an anxiety-provoking process. The field supervisor who has developed a level of trust and closeness with a student may find it difficult to get enough distance to evaluate that student objectively. He or she may want to avoid potential conflict with the student by avoiding difficult issues or less glowing aspects of the student's performance. The evaluation process might also prompt self-evaluation in the supervisor. The supervisor might begin to question his or her own methods of working with the student and wonder whether the student might have been more successful if the supervisor had done this or that differently.

The supervisor also brings to the evaluation process his or her own experiences with evaluation. The evaluation standards, processes, and methods used within the settings in which the supervisor has worked undoubtedly cast a larger context in which student evaluation is viewed. For example, a supervisor who is working or has worked in settings in which worker evaluations were rigorously conducted is much more likely to conduct rigorous evaluations of students. In other words, field supervisors bring their own humanness and experiences to the evaluation process and often have their own thoughts, feelings, and histories to sort out in relation to the evaluation process. Although every student might hope for a supervisor who will simply pat them on the back, write a consistently glowing evaluation, and send them happily on their way, most students also realize that this type of evaluation in the end is not very helpful.

The purpose of your evaluation is to help you leave your placement with a clear grasp of your current level of skill and knowledge as well as with some clear goals for your future learning, growth, and skill development. In order to achieve this purpose, your supervisor must rely upon objective, measurable, observable data about your work in completing your evaluation (Faiver et al., 1995). An evaluation that is too vague or general will not give you the feedback that you need either to recognize your own strengths or to improve your future performance. Neither does an evaluation that is too "one-sided," addressing only strengths or only weaknesses, give you this focus. With all of this in mind, expect that you and your field supervisor will identify areas of strength as well as areas that need to be improved and that these impressions will be drawn from objective observations of your work. Given that this is the expected range of focus for any evaluation, suggestions for future improvement should not be taken as criticism. Even the most experienced and proficient practitioners can identify areas of their performance that need to be improved.

To understand your supervisor's perspective, it will be helpful for you to keep in mind the issues of accountability affecting your supervisor's evaluation of you. As discussed earlier, your field supervisor has responsibilities not only to you but to other constituencies as well. He or she must evaluate you in a manner that gives consideration to the future clients and organizations whom you may serve and to the integrity of the profession as a whole. Also, your field supervisor is accountable to your academic department. An important duty of the field supervisor is to give the human services faculty objective, honest feedback about each student's abilities and learning needs. As your supervisor completes your evaluation, all of these levels of responsibility will come into play.

THE EVALUATION CONFERENCE

Most field supervisors will complete the written evaluation and share it with the student in some form. For your evaluation to have the greatest learning benefits for you, you should have the opportunity to discuss the evaluation with your supervisor. In light of the earlier discussion of how difficult evaluation can be for both the student and the supervisor, it is perhaps understandable that at times one or both parties may try to avoid this direct discussion of the evaluation. If your supervisor does not offer to discuss your evaluation with you, it is appropriate and advisable for you to request such a discussion (Chiaferi & Griffin, 1997). Useful discussion topics for this conference include any particular questions you might have about your supervisor's ratings or assessments, areas in which your perceptions and your supervisor's perceptions differ, your and your supervisor's perceptions of your greatest strengths and weaknesses, and ideas about how you might improve your performance in your next experience, whether this should be another educational fieldwork experience or a job.

A major goal for you as the student in this conference is to develop a clearer and more detailed understanding of how your work has been perceived by your supervisor.

Through this discussion, you can possibly develop a fuller and more complete picture of your performance, one that is enhanced by seeing yourself through the eyes of a more experienced professional worker. A further goal of the conference is to develop some ideas regarding how you might use your supervisor's feedback to focus and direct your future growth and development. Having said this, it should be clear that the discussion is not for the purpose of trying to modify your supervisor's evaluation or perceptions, although this sometimes occurs as a by-product of the discussion. It is appropriate for you to point out any concrete evidence you might have to support a rating different from your supervisor's (Thomlison et al., 1996).

Although the conference needs to be a conversation involving two-way communication between you and your supervisor, keep in mind that the conference will have the greatest value for you if you can enter it with a spirit of openness and flexibility. You will want to share your own perceptions honestly and straightforwardly with your supervisor, but it is also important to be ready to listen, recognizing that the conference is potentially one of the most important learning experiences of your internship. Therefore, it is wise to enter the conference with a nondefensive posture, with your mind and ears open, so that you can hear what your supervisor has to say.

A Student's Reflections on the Evaluation Conference

I was a little nervous going into my evaluation conference. I didn't think there would be any big problems, but still evaluation time is kind of like test time and it always makes me nervous. I was relieved that the conference went great! She reassured me about my goals and my career choice. Even though I know that this is the field I want, it is wonderful to hear an experienced professional say that I will be good at it. She noted particularly my strengths in working with children, my writing skills, and my ability to get work done promptly. We also discussed the fact that I need to continue developing my computer skills and my assertiveness in certain situations. Overall, I felt it was a very positive evaluation but one that did not gloss over the "needs to improve" category. I appreciated her thoroughness and honesty in discussing the evaluation with me.

Exercise 11.6

Complete this exercise after you have had your evaluation conference with your supervisor. In reflecting upon this discussion and the feedback you received, discuss the following questions: What messages were you most pleased to receive from your supervisor? What messages were most difficult for you to hear or accept? How well were you able to maintain an open and nondefensive posture throughout the discussion?

How might your supervisor's feedback clarify or reshape your goals for future improvement and growth?

If you were to have a job interview today, what particular skills and strengths could you confidently tell a potential employer that you demonstrated during your internship?

THE FACULTY LIAISON'S PERSPECTIVE

The faculty member from your academic department who works with you in your internship might be referred to by a number of titles, including the "faculty liaison," "faculty supervisor," or "internship instructor." Whatever the title, this individual has an important role in determining your final grade for your fieldwork. The faculty member, as a college employee, is responsible for assigning the final grade, taking into consideration the field supervisor's evaluation as well as your performance on other requirements of the course, such as written assignments. How various components of your performance are weighted will vary from program to program and may even vary some from one faculty member to another.

As discussed earlier, important issues of accountability will come into play as your field liaison determines your final grade. Your faculty liaison is responsible for upholding the standards of the academic department and is acutely aware that every student in the field represents both the college and the academic department. In fact, the faculty member at some point in the future probably hopes to work with your field agency and your field supervisor again as they host another student intern in the future. Each student's performance in the internship can pave the way for the next student, make the next placement rockier, or even make another placement in the agency impossible. Clearly, the academic program has much to gain as a result of those students who perform well in the field and much to lose as a result of those who do not.

The credibility and integrity of the academic department rests in part upon the program's ability to attest to the quality of its graduates. Because the field liaison works with many students in the field, he or she is perhaps in the best position to weigh the quality of students' performance in relation to one another, identifying those students who have done an outstanding job as compared to those whose work was average, minimally adequate, or unsatisfactory. In subsequent months and years, the student's grade in the internship may be included among other materials as various students are considered for graduate school admission or for employment. The faculty member's personal credibility, as well as that of the department, will be damaged if the grade does not honestly and accurately reflect the student's ability and the quality of the student's performance.

As a human services professional, the faculty liaison (like your field supervisor) is also accountable to the profession. He or she has a stake in ensuring that all those entering the field will perform adequately to meet the needs of clients as well as to meet the expectations of the employing agencies. From an ethical perspective, the faculty member might be seen as a gatekeeper of the profession, ensuring that only competent practitioners earn the credentials of the profession and recognizing that incompetent practitioners damage the profession as well as the clients they serve and the agencies that employ them.

Finally, your faculty liaison is accountable to you. Your faculty liaison wants you to be successful in your internship and no doubt would like you and every student from the program to *earn* the highest possible evaluation. You can be sure that this faculty member wants to do everything possible to support your success throughout the internship. As a faculty member in your academic department, he or she may have a long-standing relationship with you. This relationship might make objective and rigorous evaluation of your performance particularly difficult. Nevertheless, your faculty liaison's accountability to you in the evaluation process requires that he or she be honest and objective in giving you feedback. Your faculty liaison and field supervisor share responsibility for ensuring that you leave your fieldwork with an honest and realistic picture of your strengths and weaknesses as a practitioner, especially since this information should guide your continuing professional development. The responsibilities of both the teaching profession and the human service profession require the faculty member and the field supervisor to formulate honest, objective assessments of your performance and your level of competence as a developing human services professional.

CONCLUSION

A summative evaluation of your work will occur toward the end of your field placement. This evaluation process will call upon you to take an active part in assessing your learning and your performance. You, your field supervisor, and your faculty liaison will have important roles to play in your final evaluation. Issues of accountability will strongly influence the evaluation process as they do human services education and the human services profession. Similar evaluations will continue throughout your career because evaluation is an essential component of accountability within the profession and your own professional growth and development. As the content and exercises in this chapter illustrate, human service professionals should strive not only to be open to feedback from others but also to engage in continual personal reflection and honest self-evaluation. Through this process, they can evaluate their strengths and weaknesses over time, set goals for ongoing learning and growth, and monitor their own progress toward achieving these goals. The commitment to such a process is a hallmark of a true professional in any field.

Your Reflections on the Evaluation of Your Performance

REFERENCES

Alle-Corliss, L., & Alle-Corliss, R. (1998). *Human services agencies: An orientation to fieldwork.* Pacific Grove, CA: Brooks/Cole.

Barker, R. (1996). *The social work dictionary.* Washington, DC: NASW Press.

Bogo, M., & Vayda, E. (1995). *The practice of field instruction in social work: Theory and process.* Toronto: University of Toronto Press.

Chiaferi, R., & Griffin, M. (1997). *Developing fieldwork skills: A guide for human services counseling, and social work students.* Pacific Grove, CA: Brooks/Cole.

Faiver, C., Eisengart, S., & Colonna, R. (1995). *The counselor intern's handbook.* Pacific Grove, CA: Brooks/Cole.

Kadushin, A. (1976). *Supervision in social work.* New York: Columbia University Press.

Kaiser, T. (1997). *Supervisory relationships: Exploring the human element.* Pacific Grove, CA: Brooks/Cole.

Simon, E. (1999). Field practicum: Standards, criteria, supervision, and evaluation. In H. Harris & D. Maloney (Eds.), *Human services: Contemporary issues and trends* (2nd ed., pp. 79–96). Boston: Allyn & Bacon.

Thomlison, B., Rogers, G., Collins, D., & Grinnell, R. (1996). *The social work practicum: An access guide.* Itasca, IL: Peacock Publishers.

Wilson, S. (1981). *Field instruction: Techniques for supervisors.* New York: Free Press.

Leaving Your Field Placement

As you approach the end of your fieldwork, you might find leaving your internship to be every bit as challenging as beginning it was some time ago. Depending upon the nature of your internship site as well as the nature of your relationships with your supervisor, co-workers, and clients, leaving the internship may be an experience of great emotional intensity or one of relative ease (Sweitzer & King, 1999). In any case, the work of leaving a field placement of any type will focus on such tasks as ending your relationships in a positive way, reflecting upon and evaluating your experiences, examining your own growth and development, and writing any necessary closing and/or transfer summaries. In order to handle these tasks well, it is helpful to know as much as possible about the termination process. This chapter will introduce you to important guiding principles for ending your fieldwork effectively and will explore the specific tasks and processes that are necessary to bring your work to a positive close.

A Student's Reflection on Leaving the Internship

This has been the single most valuable experience I have had as a college student. I can't imagine anything that I could have done this semester that would have taught me more. I'm not ready to leave and I don't think the people at the group home (staff or clients) are ready for me to leave either. All of the children here have been left again and again throughout their lives. I feel guilty about being another one of those people. They keep asking me why I have to go, when my last day is, what I'll be doing after I leave, and so on. I guess it helps them to talk about it, but I find it difficult every time it comes up. Today I was talking with a little boy whose parents have lost their parental rights. He is waiting to be adopted. As we were talking about my leaving he suddenly asked, "Will you be my mama?" This was extremely difficult for me to handle without getting upset, but somehow I muddled through it. I am finding that leaving this internship is the most exhausting part of the entire experience.

AN OVERVIEW

Ending relationships can be difficult. The process of leaving your internship will involve ending not just one relationship, but many. As a result, this period of time can challenge your interpersonal skills and your emotions. The technical term that is generally used in discussing such endings in human services contexts is *termination*. Termination is defined as "a systematic procedure for disengaging the working relationship" (Barker, 1996, p. 380). The termination stage of the internship was briefly discussed in Chapter 1, "Getting Ready," as you were introduced to the stages of internship development. Just as the end of the internship was discussed at the beginning of this book, so your termination from the internship has been anticipated from its beginning. You, your clients, your supervisor, and your co-workers have known since the outset that your stay in the organization was time-limited. Possibly there was from the beginning a very specific predesignated date for ending your fieldwork. In spite of this, the time to leave your internship can seem to come upon you suddenly and before you are ready for it. Often, students express resistance to ending with comments such as, "I was just beginning to get good at this. I don't want to leave now!" At the same time, students often express great pleasure, pride, and satisfaction in ending their fieldwork due to the rich and valuable experiences that they will take with them and their sense of having performed a job well. Feelings of accomplishment and pride are often mingled with feelings of sadness, regret, and even relief as the experience draws to a close. Such reflections and mixed feelings are predictable and normal aspects of the termination process.

Conducting a positive termination process will call upon you to engage in extensive reflection on your experiences over the course of the internship. Your formal performance evaluation (discussed at length in the previous chapter) is one aspect of this reflection. Evaluation of your field placement and your field supervisor are also required within most field programs and are important steps in thoroughly processing your experience. Similarly, you will benefit from assessing your own progress and beginning to think about the next steps in your own learning. In preparing to leave your agency, you will also need to make careful plans for each of the clients and/or projects that need continued attention after your departure. Throughout the termination process, you will also be involved in the sometimes difficult emotional work of saying good-bye, allowing yourself and others to feel and express whatever emotions are present as you go about the business of leaving. Understanding the principles involved in termination will enable you to leave your internship in a manner that will enhance your learning and bring closure to your work.

GENERAL GUIDELINES FOR POSITIVE TERMINATION

In your academic coursework, you have probably studied how to terminate with clients effectively. Reviewing any coursework you might have done on this topic or seeking out relevant reading would be a good idea as you begin terminating from your internship. Many of the principles that guide termination of the helping relationship will be helpful to you in terminating from your internship. The following principles of positive termination from your internship are adapted from guidelines for handling termination within helping relationships. You are more likely to have a positive termination from the internship if you follow these principles, not only in ending your relationships with your clients but in working through your own termination from the internship as well.

Be Aware of Your Previous Experiences and Patterns with Terminations (Patterson & Eisenberg, 1983)

All of us have dealt with losses. In the course of normal life there are many relationships and situations that ultimately end in separation and saying good-bye. Some of these events are intense and difficult, while others are easier, perhaps just a normal part of growing up. Moving away from friends and family, changing schools, breaking up with a romantic partner, and experiencing a divorce or the death of a loved one are all common life events that people frequently have to deal with. Over the course of dealing with such separations, individuals tend to develop patterns in their reactions, feelings, and behaviors surrounding separations.

Identifying and evaluating your own patterns in dealing with endings is a good first step in preparing to handle the many terminations inherent in leaving your internship. Likewise, it is helpful during this time to engage clients in similar exploration, helping them to identify their own patterns in saying good-bye. Clients who can approach termination with this degree of self-awareness can often feel more in control simply because they can better anticipate the feelings and reactions that they are likely to experience along the way.

A Student's Reflections on Termination Patterns

In my personal life I generally try to avoid good-byes. Whenever I've moved or left friends behind, I always kind of hide behind a "see you later" type attitude. I make lots of promises to keep in touch and the other people do the same, but it usually doesn't happen. I will be graduating from college soon, and already I am saying these same lame lines to my friends. Of course, true to form, this is also my tendency in my internship as well. It is really tempting to say I'll keep in touch, come by to visit, and so on. I realize that this is where I have to draw the line with my old patterns. I can't make easy, false promises here. I need to be honest and say a real good-bye. I hate doing that because it is so sad.

Exercise 12.1

Identify a few particularly meaningful separations that you have experienced previously. What were your emotional reactions to these separations? What were your behavioral reactions? What strategies did you use to deal with the losses involved in these separations?

How might these patterns be helpful or unhelpful as you terminate your internship? In what ways might you need to alter these patterns in order to terminate from your field placement in a positive and healthy manner?

Be Self-Aware, Recognizing Your Needs and Wants (Patterson & Eisenberg, 1983; Sweitzer & King, 1999)

A range of reactions is normal at the time of termination. Being aware of these reactions is a necessary step in managing terminations effectively. In terminating with clients, they need to be encouraged to recognize and express their wants and needs as they anticipate the end of the relationship. These wants and needs may not be fixed and static, as clients often express different reactions at various points in the termination process or even simultaneously. At times, their wants and needs might even be contradictory and conflicting. Such ambivalence (a state of mixed feelings and reactions) is a normal response to termination (Chiaferi & Griffin, 1997). Because ambivalence can create feelings of confusion and distress, it is especially important to explore such feelings and offer emotional support.

Just as clients benefit from addressing their needs and wants at the time of termination, you can do likewise. At times, you might be eager to terminate, ready to move on to the next challenge in your life or ready to take a break. At other times, you may feel resistant to leaving your placement. You might feel that you have not yet learned enough, or you might regret that you have not been as involved as you could have been. Being aware of your wants and needs can help you to avoid some common pitfalls in termination, such as "premature disengagement" from your internship (discussed in Chapter 1) or denial of the pending termination. Students who disengage prematurely from the internship tend to withdraw from clients and co-workers and may even become less attentive to their work responsibilities as the end of their internship approaches. Students who deny the pending termination may fail to give clients adequate warning of their leaving or may give clients false reassurance by promising that they will keep in touch. Therefore, it is important to be honest with yourself about your wants and needs regarding leaving your internship so that you can avoid such pitfalls in your work. During termination, students, like clients, can also feel confused and ambivalent about what they want and need.

A Student's Reflections on Wants and Needs

At times I find myself desperately wanting a job here. At these times I simply don't want to leave. The fact that there is some slim chance that I might actually get a job here within the next several months makes it all the more difficult to let go. At other times I am ready to go. I think the main reason I feel this way at times is that I am just very tired. I need a break. I am very grateful for the opportunity I have been given here, but I do feel like it is time to stop being a student. As much as I've liked my internship, it is a little bit like "Never-land." I'm not a kid and not a grown-up, not an employee and not a volunteer. I want to be a full-fledged, employed, paid professional somewhere, and most of the time I would like that place to be here.

Exercise 12.2

As you approach the end of your internship, what are your wants and needs? In what ways are you ready to end the internship? In what ways are you resistant to leaving?

Reflect Upon and Deal with Your Feelings (Alle-Corliss & Alle-Corliss, 1998; Brill, 1995; Brown & Pate, 1983; Sweitzer & King, 1999)

Clients need to be encouraged to express their emotions about the approaching end of the helping relationship. Emotions around termination can cover a wide range from sadness, anger, resentment, and frustration to relief, joy, satisfaction, and pride. For both clients and yourself, some of these feelings may be easier to acknowledge and discuss than others. Give yourself and others the permission to experience the full range of emotions that may be present.

Whatever your feelings are, it is helpful to verbalize them to someone who is a good, accepting listener. Your field supervisor may want to explore your feelings about termination with you. Also, discussing your feelings with other internship students is often very helpful because you are all going through similar experiences. The field seminar is an ideal opportunity for this type of discussion. Discussing your feelings is an effective way of becoming more comfortable with them and bringing emotional closure to your internship.

Exercise 12.3

What emotions are you experiencing as you anticipate the end of your internship? Which of these emotions are you most comfortable with? What emotions, if any, are you uncomfortable acknowledging or discussing?

Review the Experience (Alle-Corliss & Alle-Corliss, 1998; Patterson & Eisenberg, 1983; Young, 1998)

As clients engage in the termination process, they are invited to review the helping process, identifying the critical events that led to positive change for them. These events are sometimes thought of as turning points in the client's growth and development. It is not uncommon for clients at the end of the helping relationship to look back at some of the most difficult or discouraging points of the process as being most important in retrospect.

As you have experienced your internship day-to-day, you have probably been aware of events that seemed particularly exciting, interesting, or educational. At other times, you might have found particular events to be tedious, boring, stressful, or even distasteful. In the immediacy of the everyday demands, it can be difficult to recognize the value of particular events. Toward the end of your placement, as you reflect on your experience as a whole, you are in a better position to identify its most valuable components. Surprisingly, some of the most valuable events of the internship might be those that at the time were unpleasant or seemed to be of little consequence. Conversely, those events that seemed very significant or positive at the time might over the long run prove not to be so important to your overall development and learning.

A Student's Review of Key Events

It is really ironic that I chose this internship because it dealt with juvenile crime. In fact, this was about the only thing I even knew about the placement before I started it. As it turns out, this has probably been its least important feature. What I have gotten interested in and learned about here has been family dynamics and parenting issues. The problem focus could have been just about anything. What I have gotten out of this has been so much broader and deeper than just learning about a single population or problem. It has opened my eyes to a whole new way of understanding human behavior. My goals have changed dramatically as a result. I am exploring graduate programs in family counseling now.

Exercise 12.4

A key question to consider as you reflect upon your internship is, "What particular events in or components of this internship have caused me to learn, grow, and change?" Identify at least three specific events in your internship that have been particularly meaningful in your experience. In the case of each event, try to articulate what gives the event special significance.

Acknowledge the Progress and the Changes That You Have Made (Cormier & Hackney, 1993; Patterson & Eisenberg, 1983; Royse et al., 1999)

When clients terminate from a helping relationship, it is especially helpful to reflect with them upon the gains they have made during the course of the helping relationship. Acknowledging these gains, being able to name them and claim them, helps clients to retain their progress once the relationship has ended. Furthermore, recognizing their gains enables clients to leave the helping relationship with a sense of pride and accomplishment. A similar process will be helpful to you as you leave your internship. It is important that you think about your development over the course of the internship, focusing particularly on what you have learned and how you have grown and changed, professionally and personally. As with clients, recognizing and acknowledging the progress that you have made in your own growth and development will give you a firmer grasp of what you have learned and a sense of pride and accomplishment.

A Student's Reflections on Accomplishments

I divide what I've actually achieved in this internship into two categories: what I achieved for myself and what I achieved for the agency. For myself, I feel that I have made the big shift into professional life. I've learned to hold my own in a group of professionals, to carry my own workload competently, to handle difficult people, and to juggle many tasks. I have learned how to work on multiple projects, with multiple teams, and multiple bosses. I know that I can go into a job interview now and talk about my abilities in a more convincing way because now I really know that they are there. For the agency, I have organized a county-wide agency fair (almost single-handedly), written a small grant for some playground equipment, and led a major fund-raising event. In addition, I hope and believe that I have offered some support and encouragement to a number of children who really needed it. Have I changed the world? No. Have I made a difference? I can say with confidence, "Yes!"

To fully appreciate what you have accomplished during your internship, it might be helpful to think about your first days there and compare them to your experience now. In doing so, it might also be helpful to review some of your writing in response to earlier exercises in this book. Such a review will probably reveal some fairly significant learning and growth over the course of your internship.

418 Rose St.
5703 Hwy 220
1320 PA
1354
PA.

Exercise 12.5

As you look at your development over the course of the internship, what are your observations? In what specific ways have you grown, learned, and changed during the internship?

Imagine that you are interviewing for a human services job or for your next field placement and that the interviewer asks, "So what did you get out of that internship?" How would you respond to such a question?

Anticipate and Plan for Future Needs (Brill, 1995; Brown & Pate, 1983; Cormier & Hackney, 1993; Royse et al., 1999)

In terminating with clients, the worker helps the client to think about the future. Together, worker and client anticipate the concerns and needs that might arise and plan how these future needs might best be met. As they anticipate together the challenges that the client might face in the future, they consider how newly acquired skills might be applied to these situations as well as anticipate any needs for additional resources and support that the client is likely to encounter in the future.

In applying this principle to yourself, think about your future and your continued learning needs. Students often leave internships with a great appreciation for how much they have learned as well as for how much they *don't* know. As you enter your next field placement or as you embark on your career, it is important to recognize the knowledge and skills that you currently possess, but it is also important to recognize the limits of your current abilities. Awareness of your learning needs will give focus and direction to your future efforts toward professional growth.

Exercise 12.6

As you leave this internship, assess what the next steps might be in your continued learning. What knowledge and skills do you need to continue developing as you enter your next human services experience? Identify at least three to five goals for your own continued learning as well as a few strategies for achieving these goals.

These guiding principles for termination emphasize the importance of intensive review, reflection, and evaluation as you leave your field placement. Keep these principles in mind, remembering to apply them to your clients as well as to yourself, as you go about the tasks involved in termination. These principles will be particularly helpful to you as you bring the most meaningful relationships of your internship to a close.

SAYING GOOD-BYE TO YOUR SUPERVISOR

Although you might have worked with many professionals during your internship, your field supervisor probably holds special significance in your experience. This individual, more than any other, generally has the greatest impact on student experiences in the field. It has been argued, in fact, that the student-supervisor relationship is of primary importance in student learning, offering the best medium for preparing students to meet the demands of practice situations (Matorin, Monaco, & Kerson, 1994). Because of its significance, your relationship with your supervisor should be terminated in a manner that respects the general principles for positive termination which were discussed earlier in the chapter. Many supervisors ensure that this occurs by initiating discussions in which the high points and low points of the supervisory relationship are explored, the student's growth and development within that relationship are evaluated, and feelings about ending the internship are expressed. Such discussions with your supervisor are beneficial for you as an intern in two ways. First, these discussions provide a structured time and opportunity in which you can explore and work through the process of termination. Second, such discussions offer you the opportunity to see your supervisor modeling the steps and professional skills that are involved in effective termination (Bogo & Vayda, 1995).

As you and your supervisor reflect on your internship experience, issues related to the evaluation of your supervisor and of the agency as a placement can sometimes surface (Sweitzer & King, 1999). Your academic program will most likely require you to complete a formal evaluation of your field placement and/or your supervisor. This usually involves your completion of an evaluation form so that systematic information can be gathered by your program about all of the various field sites that host student internships. In addition, your field supervisor might ask you for feedback about the placement. Many field supervisors want to know how they could improve the experience as well as what components of the experience were particularly valuable. Although not all supervisors will ask for such feedback, those who do generally are quite sincere in wanting your honest assessment. This can be, however, an awkward experience for you as a student. If you should be in this position, try to remember that just as you cannot learn from your evaluation unless it is shared and discussed with you, field supervisors cannot benefit from feedback that is not shared with them.

Understandably, giving your supervisor feedback might be difficult for you to do in certain circumstances. If you feel, for example, that your placement did not fully meet your learning needs due to the nature of the agency's mission, this can hardly be attributed to your supervisor's handling of your internship. Therefore, such feedback will probably not be too difficult for you to share. If, however, you feel that your supervisor withheld certain types of experiences from you that were potentially available, this might be more difficult to address. Nevertheless, such issues are important to discuss. In such a discussion, your supervisor might be able to point out particular aspects of your academic preparation or internship performance that suggested you were not ready for these experiences. This feedback could be enormously helpful to you as you prepare for your next field placement or for employment. By the same token, your feedback might prompt your supervisor to reevaluate the duties assigned to you and possibly modify them for the next student who enters the agency. In either case, the conversation will have yielded important benefits. Although it may be difficult to acknowledge the less positive aspects of your placement, it is beneficial to discuss them so that both you and your supervisor can learn from these events.

Because you are the student, more accustomed to getting feedback from your supervisor than giving it, you will probably feel the need to think carefully as you prepare for this conversation, identifying clearly the issues you would like to address and how to express yourself tactfully. Just as your expectation was that your supervisor would recognize your strengths as well as your shortcomings as a student, you would do well to strive toward the same balance and fairness of mind in your comments. Some field supervisors and placements, like some students, might have been of such high quality that it is indeed difficult to make suggestions for improvement. Even so, students in such placements should try to identify some ways in which the supervisor and the agency could further improve the quality of their work with students.

Exercise 12.7

Identify at least two of your supervisor's most important strengths in his or her role as an internship supervisor. To identify these, you might think of the qualities that you particularly appreciated about your supervisor, qualities from which you benefited as a student. Also identify at least two specific suggestions as to how your supervisor might have further enhanced your learning experience.

In a similar fashion, identify at least two particular strengths of the agency as a field site for students. Also make at least two suggestions as to how the agency could improve its work with students in the future.

SAYING GOOD-BYE TO YOUR CLIENTS

Most field placements involve at least some direct work with clients. Whether these clients have been children in a child care center, residents in a group home, older adults in a senior center, inmates in a prison, employees in a company, or students in a school, you have come to mean something to those with whom you have worked, and they have come to mean something to you. If your field placement has been more administrative in nature, you have perhaps had less contact with traditional client groups and have instead worked more closely with staff or served other populations. These relationships, however, may be equally meaningful and difficult to leave behind. Therefore, all concerned, those with whom you have worked as well as you yourself, will need an adequate opportunity to prepare for your leaving.

For many students, one of the most difficult aspects of leaving the internship is terminating with clients. The process will be much smoother, of course, if you have been clear with your clients from the outset that you were a student in a time-limited placement. Clients need sufficient warning that the end of your relationship is approaching and a clear date as to when your last contact will occur. It is often helpful to taper off your relationship by gradually reducing the frequency and/or the length of your contacts. Such a process can help to reduce client anxiety about termination (Okun, 1992). For some clients, such a transition might not be warranted. These clients might continue working with you until the end of your internship, then transfer to another worker at that time. In this case, the client will still need a chance to terminate with you and to adjust to the idea of a new worker. In any case, you need to introduce the idea of termination early enough to make whatever transition is necessary in the client's care and to ensure that the client does not feel that your relationship was abruptly cut off.

In some settings, such as residential programs and group treatment settings, you might have worked as part of a treatment team without responsibility for ongoing, one-to-one contacts with clients. In such cases, there will probably not be a designated worker to replace you. Clients in such settings will not be transferred to new workers and will benefit from the continuity of the other team members' ongoing availability. Nevertheless, worker-client relationships within such settings can become quite close and meaningful. Consequently, your clients will need the opportunity to engage in the termination process with you as you leave your internship. Upon your departure, remaining team members might be asked to provide additional informal support for those clients whom you anticipate having the most difficulty with your leaving.

A Student's Reflections on Saying Good-bye to Clients

Since my internship has been in a senior center, I honestly didn't think my leaving would be a big deal. A lot of the termination stuff we have studied I thought was mostly relevant to casework or counseling situations. Much to my surprise, there have been lots of different reactions to my leaving. One of the ladies got teary-eyed a few days ago when we were talking about my last day. Mr. T., whom I have been playing cards with almost every day, seems to be avoiding me now. They all tease me about leaving, saying things like, "Go ahead and just abandon all us old folks!" I know they are teasing, but I still feel bad when they say it. I also know, as my mother always said, that "many a truth is said in jest." I'm going to talk to my supervisor about having a party or something before I leave. I think we need to do something to make it easier.

Exercise 12.8

How well prepared are your clients for the end of your internship? To what extent have they been aware throughout your internship that you would be leaving at a specific point in time? Are there particular individuals for whom you feel your leaving might be especially difficult? If so, how might you make a special effort to prepare them at this point?

In terminating with clients, there are several tasks to accomplish. All of these tasks flow from the guiding principles for termination discussed earlier in the chapter. As suggested there, you should engage clients in reflecting upon their work or relationship with you. This reflection should include evaluating their progress and identifying key events that enabled them to make progress. In addition, clients should be encouraged to discuss their feelings about termination and to make any plans for the future that are needed to support their continued stability and growth.

A complicating factor in terminating with clients is the fact that each individual has his or her own history of losses and his or her own patterns in dealing with those losses. As a result, you might find that no two clients react in exactly the same way to your leaving. Just as you explored your own history and patterns in saying good-bye earlier in the chapter, it will be useful to explore possible reactions from clients as well. Although some clients may have intense emotional reactions to termination, most evidence suggests that client reactions are generally positive. Clients are much more likely to have positive reactions to termination if the ending has been anticipated and planned for systematically (Fortune, 1995). Clearly, the worker can do a great deal to ensure that termination is handled in this manner. Nevertheless, Barker states that terminating with clients often prompts emotional upheaval and cites client resistance, denial, and flight into illness as common client reactions (1996, p. 380). Each of these responses will be briefly considered.

Resistance occurs when clients try to avoid or impede the worker's efforts to discuss termination. Clients might resist, for example, by forgetting that you are leaving or forgetting the date of your departure. They might change the subject when the topic of termination is brought up. Likewise, the resistant client might fail to show up for appointments, particularly your last contact or the appointment in which the new worker is to be introduced. In such cases, it might be useful to point out to the client the pattern that you see and to discuss his or her patterns of saying good-bye in previous situations. Empathizing with how difficult it can be to let go and say good-bye can help give the client the emotional support needed to confront the situation. Resistance may be particularly intense in relationships that have been especially close and important to the client.

In contrast, some terminations may be marked by a lack of emotional intensity. This might be due to a lack of intensity in the relationship or due to client denial. For example, a minimal reaction could be predicted and is understandable from clients with whom you have had little contact or those with whom your work has been fairly limited. Many helping relationships, however, are characterized by personal disclosure and trust that have developed over a significant period of time. In such relationships, more emotional reactions from clients are to be expected. Therefore, it is significant when in such a context a client insists upon having no feelings about ending the relationship. Such clients may be denying their genuine feelings as a defense against potential emotional pain. In such a situation, it might be useful for you to express your own feelings of sadness and other genuine feelings about saying good-bye. Of course, it is important to keep this self-disclosure within appropriate limits. Keep in mind that your comments are designed to be helpful to the client and are not for the purpose of venting your own feelings or gaining emotional support for yourself (Cournoyer, 1996). Also, discussing with the client the range of emotions that clients often experience about termination can help normalize those feelings, possibly decreasing defensiveness and increasing the likelihood of more open discussion.

Another possible reaction to pending termination is sometimes referred to as *flight into illness*. This phenomenon occurs when the client suddenly begins to introduce new problems, relapse into old problems, or report a decrease in coping ability as the helping relationship is coming to an end. This reaction seems to be an effort to reengage the worker in the helping process and may be a sign of overdependency in the relationship or a fear of abandonment on the part of the client (Barker, 1996). In some cases, it seems likely that the client's increased anxiety in the face of termination might in fact create some temporary "new problems" or reduced coping ability. In any case, discussing such "relapses" with clients as a possible reaction to the pending termination can often help to allay their fears as well as help them to better understand their own feelings and behavior.

Knowledge of the three patterns discussed above—resistance, denial, and flight into illness—will help you to understand some of the varied client reactions that you might encounter toward the end of your fieldwork. Such reactions may present special challenges in your efforts to terminate with particular clients. Many terminations will be far more straightforward than these as clients directly express appropriate and pre-dictable feelings regarding termination. Even these more straightforward situations can be challenging for you as clients' emotional reactions can trigger your own feelings of sadness, guilt, and regret. Common client reactions to termination include feelings of sadness, anger, loss, abandonment, and/or betrayal. Those clients who have had multi-ple, difficult losses might find termination particularly painful and emotionally intense, even if they have not had a very close relationship with you. Such a reaction is some-times referred to as *transference*, because the client seems to transfer feelings from previ-ous relationships onto the relationship with the helper. Whatever the nature and source of the client's feelings about termination, your role is to convey your empathy and ac-ceptance of those feelings and to encourage the client to express and explore whatever feelings are present. This process of emotional expression is an important part of bring-ing your relationships with clients to a positive end.

Exercise 12.9

What emotions and reactions have you observed among the clients you serve as your termination approaches? How have these feelings been conveyed verbally and nonver-bally? In what instances, if any, have you been able to observe signs of denial, resis-tance, flight into illness, and/or any other problematic patterns?

In addition to the evaluative, reflective, and emotional aspects of termination, the more concrete tasks of closure must also be accomplished. Among the more concrete tasks is that of transferring any unfinished projects and/or ongoing clients to other workers.

TRANSFERRING YOUR WORK

Much of the work that you were assigned during you fieldwork will most likely be completed by the time you leave the organization. It is not uncommon, however, for students to have to leave some of their work unfinished. Work with clients is especially likely to extend beyond the time frame of your field placement. A most important aspect of terminating from your fieldwork is ensuring that a smooth transition is made in any unfinished work as you leave the organization. Tasks and projects that are incomplete as well as clients who need ongoing assistance must be transferred to staff members who can bring the work to completion. You and your supervisor will plan well in advance which clients and projects are likely to need ongoing attention after you leave, thus requiring transfers to new workers. For those clients who will be continuing with new workers, it would be helpful for you to write transfer summaries to ensure that the new worker has all the information necessary to provide continuity of care. Transfer summaries include content similar to that in closing summaries (see Chapter 8) but serve the purpose of summarizing your work with the client so that the case can be smoothly transferred to a different worker for continued services. Also, it is best that you introduce clients to their new workers if at all possible before you leave. These introductions might be followed by a short meeting of the three of you together. This meeting should focus on a brief summary of your work together and a general overview of the issues that still need to be addressed. Most important, the meeting is an opportunity for the client and the new worker to begin building a relationship. Your role in this meeting is to support and encourage this new relationship, providing a smooth transition for the client from one worker to another.

As this chapter has emphasized, leaving your field site can be emotionally difficult. Many students find it especially difficult when they have to leave important components of their work midstream. Unresolved issues with clients and unfinished administrative projects can add to a student's reluctance to end the placement and create the feeling of lack of closure. In such situations, students may be especially tempted to maintain their relationship with the agency in order to see certain cases and/or projects through to completion. If your thoughts are leaning in this direction, discuss them with your field supervisor and faculty liaison. Consider carefully your motives for staying involved. Although in some cases it may be appropriate to extend the placement for a short period of time in order to finish a certain specific task, in most cases it is best for students to make a clear termination at the appointed time. The nature of human services work is such that it is often difficult to bring every task to complete closure by any particular predesignated date. Therefore it is sometimes necessary to let go of your work, passing it to other workers to be brought to a close.

Exercise 12.10

What has your emotional reaction been to letting go of your responsibilities in the agency and transferring your responsibilities to others? What particular concerns or difficulties are you having in letting go of certain clients or responsibilities?

Are you tempted to try to stay in touch with any particular clients or to continue with any particular projects? If so, why?

What are your thoughts about the appropriateness of any continued contact following the end of your internship? Discuss your thoughts about this with your supervisor.

TERMINATION RITUALS

Rituals are a common and effective part of saying good-bye in many cultures as well as in many human service agencies and other types of organizations and groups. Often, rituals seem to provide comfort because they offer structure during emotionally intense situations (Fortune, 1995). Whether you have worked with groups of clients or individuals, it is likely that there will be certain ritualistic components of your leave-taking. A client might give you a small gift, a child might make you a going-away card, a group that you have co-led might mark your leaving with a going-away party, or your co-workers might take you to lunch. Such events are important ways of communicating feelings and bringing a more ceremonial or celebratory close to the experience. Rituals can be spontaneous and brief, as when a client quickly gives you a word of thanks and a hug before walking out the door, or they can be lengthy and planned, as when a group plans a potluck lunch as a going-away surprise. You, too, might want to introduce some simple elements of ritual into your ending. Writing thank-you notes or finding other ways of expressing appreciation, for example, might be especially appropriate rituals as you leave your co-workers and supervisor. By the same token, brief good-bye notes to clients in which you recognize their special strengths and express your pleasure in having worked with them might also be appropriate. In any case, you might find that rituals, initiated by yourself and/or others, bring greater depth and meaning to your termination experience as well as a greater sense of closure as you say good-bye.

A Student's Reflections on Rituals

My last day of group my supervisor did a really neat thing. We focused a lot on the fact that this was my last time with the group. He used the Gestalt idea of having everyone say good-bye to me by expressing their appreciations and regrets. I also said good-bye to all of them and to him by expressing appreciations and regrets. Everyone got a piece of drawing paper and crayons. They were asked to draw a going-away gift they would like to give me. One boy gave me a new car because he has seen what a rusted-out heap mine is. Another boy drew me a picture of himself. My supervisor planned this because a lot of the guys have trouble talking about their feelings. Most of them have had a lot of people come and go in their lives without any way of talking about it. He and I also got to model for them that guys can talk about their feelings, and it's OK. I'll have to say that even though this was planned mainly because it was good for them, it was also good for me. It was one of the most memorable experiences of my entire internship.

Exercise 12.11

What is your reaction to the idea of using rituals in terminating with clients and/or co-workers in your internship?

What rituals, if any, are being planned to mark your leaving the agency?

What kinds of rituals would you find meaningful and appropriate? Are there any rituals that you would like to initiate as part of your saying good-bye to clients and/or co-workers?

CONCLUSION

Leaving your internship involves many of the same processes as terminating a professional helping relationship. Evaluation of the experience, reflection upon its most significant components, and working through the emotions involved in saying good-bye to significant relationships are important parts of this process. As you work on terminating productively from your internship, you will be working simultaneously on a parallel process of helping your clients terminate their relationships with you. The content and exercises of this chapter have been directed toward helping you to better understand the termination process and to gain insight into how to apply its guiding principles to the specific tasks involved in leaving your internship.

Your Reflections on Leaving Your Internship

REFERENCES

Alle-Corliss, L., & Alle-Corliss, R. (1998). *Human service agencies: An orientation to fieldwork.* Pacific Grove, CA: Brooks/Cole.

Barker, R. (1996). *The social work dictionary* (3rd ed.). Washington, DC: NASW Press.

Bogo, M., & Vayda, E. (1995). *The practice of field instruction in social work: Theory and process.* Buffalo, NY: University of Toronto Press.

Brill, N. (1995). *Working with people: The helping process* (5th ed.). White Plains, NY: Longman.

Brown, J., & Pate, R. (1983). *Being a counselor: Directions and challenges.* Belmont, CA: Brooks/Cole.

Chiaferi, R., & Griffin, M. (1997). *Developing fieldwork skills: A guide for human services, counseling, and social work students.* Pacific Grove, CA: Brooks/Cole.

Cormier, L., & Hackney, H. (1993). *The professional counselor: A process guide to helping* (2nd ed.). Needham Heights, MA: Allyn & Bacon.

Cournoyer, B. (1996). *The social work skills workbook.* Pacific Grove, CA: Brooks/Cole.

Fortune, A. (1995). Termination in direct practice. In R. Edwards & J. Hopps (Eds.), *Encyclopedia of social work* (pp. 2398–2404). Washington, DC: NASW Press.

Matorin, S., Monaco, G., & Kerson, T. (1994). Field instruction in a psychiatric setting. In T. Kerson (Ed.), *Field instruction in social work settings* (pp. 159–180). New York: Haworth Press.

Okun, B. (1992). *Effective helping: Interviewing and counseling techniques* (4th ed.). Pacific Grove, CA: Brooks/Cole.

Patterson, L., & Eisenberg, S. (1983). *The counseling process* (3rd ed.). Boston: Houghton Mifflin.

Sweitzer, H., & King, M. (1999). *The successful internship: Transformation and Empowerment.* Pacific Grove, CA: Brooks/Cole.

Young, M. (1998). *Learning the art of helping: Building blocks and techniques.* Upper Saddle River, NJ: Prentice-Hall.

Ethical Standards of Human Service Professionals

National Organization for Human Service Education

PREAMBLE

Human services is a profession developing in response to and in anticipation of the direction of human needs and human problems in the late twentieth century. Characterized particularly by an appreciation of human beings in all their diversity, human services offers assistance to its clients within the context of their community and environment. Human service professionals, regardless of whether they are students, faculty, or practitioners, promote and encourage the unique values and characteristics of human services. In so doing, human service professionals uphold the integrity and ethics of the profession, partake in constructive criticism of the profession, promote client and community well-being, and enhance their own professional growth.

The ethical guidelines presented are a set of standards of conduct that the human service professional considers in ethical and professional decision-making. It is hoped that these guidelines will be of assistance when the human service professional is challenged by difficult ethical dilemmas. Although ethical codes are not legal documents, they may be used to assist in the adjudication of issues related to ethical human service behavior.

Human service professionals function in many ways and carry out many roles. They enter into professional-client relationships with individuals, families, groups, and communities, who are all referred to as "clients" in these standards. Among their roles are caregiver, case manager, broker, teacher/educator, behavior changer, consultant, outreach professional, mobilizer, advocate, community planner, community change organizer, evaluator, and administrator [1]. The following standards are written with these multifaceted roles in mind.

THE HUMAN SERVICE PROFESSIONAL'S RESPONSIBILITY TO CLIENTS

Statement 1. Human service professionals negotiate with clients the purpose, goals, and nature of the helping relationship prior to its onset, as well as inform clients of the limitations of the proposed relationship.

Statement 2. Human service professionals respect the integrity and welfare of the client at all times. Each client is treated with respect, acceptance, and dignity.

Statement 3. Human service professionals protect the client's right to privacy and confidentiality except when such confidentiality would cause harm to the client or to others, when agency guidelines state otherwise, or under other stated conditions (e.g., local, state, or federal laws). Professionals inform clients of the limits of confidentiality prior to the onset of the helping relationship.

Statement 4. If it is suspected that danger or harm may occur to the client or to others as a result of a client's behavior, the human service professional acts in an appropriate and professional manner to protect the safety of those individuals. This may involve seeking consultation, supervision, and/or breaking the confidentiality of the relationship.

Statement 5. Human service professionals protect the integrity, safety, and security of client records. All written client information that is shared with other professionals, except in the course of professional supervision, must have the client's prior written consent.

Statement 6. Human service professionals are aware that in their relationships with clients, power and status are unequal. Therefore they recognize that dual or multiple relationships may increase the risk of harm to, or exploitation of, clients, and may impair their professional judgment. However, in some communities and situations, it may not be feasible to avoid social or other nonprofessional contact with clients. Human service professionals support the trust implicit in the helping relationship by avoiding dual relationships that may impair professional judgment, increase the risk of harm to clients, or lead to exploitation.

Statement 7. Sexual relationships with current clients are not considered to be in the best interest of the client and are prohibited. Sexual relationships with previous clients are considered dual relationships and are addressed in Statement 6.

Statement 8. The client's right to self-determination is protected by human service professionals. They recognize the client's right to receive or refuse services.

Statement 9. Human service professionals recognize and build on client strengths.

THE HUMAN SERVICE PROFESSIONAL'S RESPONSIBILITY TO THE COMMUNITY AND SOCIETY

Statement 10. Human service professionals are aware of local, state, and federal laws. They advocate for change in regulations and statutes when such legislation conflicts with ethical guidelines and/or client rights. Where laws are harmful to individuals, groups, or communities, human service professionals consider the conflict between the values of obeying the law and the values of serving people and may decide to initiate social action.

Statement 11. Human service professionals keep informed about current social issues as they affect the client and the community. They share that information with clients, groups, and community as part of their work.

Statement 12. Human service professionals understand the complex interaction between individuals, their families, the communities in which they live, and society.

Statement 13. Human service professionals act as advocates in addressing unmet client and community needs. Human service professionals provide a mechanism for identifying unmet client needs, calling attention to these needs, and assisting in planning and mobilizing to advocate for those needs at the local community level.

Statement 14. Human service professionals represent their qualifications to the public accurately.

Statement 15. Human service professionals describe the effectiveness of programs, treatments, and/or techniques accurately.

Statement 16. Human service professionals advocate for the rights of all members of society, particularly those who are members of minorities and groups at which discriminatory practices have historically been directed.

Statement 17. Human service professionals provide services without discrimination or preference based on age, ethnicity, culture, race, disability, gender, religion, sexual orientation, or socioeconomic status.

Statement 18. Human service professionals are knowledgeable about the cultures and communities within which they practice. They are aware of multiculturalism in society and its impact on the community as well as individuals within the community. They respect individuals and groups, their cultures and beliefs.

Statement 19. Human service professionals are aware of their own cultural backgrounds, beliefs, and values, recognizing the potential for impact on their relationships with others.

Statement 20. Human service professionals are aware of sociopolitical issues that differentially affect clients from diverse backgrounds.

Statement 21. Human service professionals seek the training, experience, education and supervision necessary to ensure their effectiveness in working with culturally diverse client populations.

THE HUMAN SERVICE PROFESSIONAL'S RESPONSIBILITY TO COLLEAGUES

Statement 22. Human service professionals avoid duplicating another professional's helping relationship with a client. They consult with other professionals who are assisting the client in a different type of relationship when it is in the best interest of the client to do so.

Statement 23. When a human service professional has a conflict with a colleague, he or she first seeks out that colleague in an attempt to manage the problem. If necessary the professional then seeks the assistance of supervisors, consultants or other professionals in efforts to manage the problem.

Statement 24. Human service professionals respond appropriately to unethical behavior of colleagues. Usually this means initially talking directly with the colleague and, if no resolution is forthcoming, reporting the colleague's behavior to supervisory or administrative staff and/or to the professional organizations to which the colleague belongs.

Statement 25. All consultations between human service professionals are kept confidential unless to do so would result in harm to clients or communities.

THE HUMAN SERVICE PROFESSIONAL'S RESPONSIBILITY TO THE PROFESSION

Statement 26. Human service professionals know the limit and scope of their professional knowledge and offer services only within their knowledge and skill base.

Statement 27. Human service professionals seek appropriate consultation and supervision to assist in decision-making when there are legal, ethical, or other dilemmas.

Statement 28. Human service professionals act with integrity, honesty, genuineness, and objectivity.

Statement 29. Human service professionals promote cooperation among related disciplines (e.g., psychology, counseling, social work, nursing, family and consumer sciences, medicine, education) to foster professional growth and interests within the various fields.

Statement 30. Human service professionals promote the continuing development of their profession. They encourage membership in professional associations, support research endeavors, foster educational advancement, advocate for appropriate legislative actions, and participate in other related professional activities.

Statement 31. Human service professionals continually seek out new and effective approaches to enhance their professional abilities.

THE HUMAN SERVICE PROFESSIONAL'S RESPONSIBILITY TO EMPLOYERS

Statement 32. Human service professionals adhere to commitments made to their employer.

Statement 33. Human service professionals participate in efforts to establish and maintain employment conditions that are conducive to high quality client services. They assist in evaluating the effectiveness of the agency through reliable and valid assessment measures.

Statement 34. When a conflict arises between fulfilling the responsibility to the employer and the responsibility to the client, human service professionals advise both of the conflict and work conjointly with all involved to manage conflict.

THE HUMAN SERVICE PROFESSIONAL'S RESPONSIBILITY TO SELF

Statement 35. Human service professionals strive to personify those characteristics typically associated with the profession (e.g., accountability, respect for others, genuineness, empathy, pragmatism).

Statement 36. Human service professionals foster self-awareness and personal growth in themselves. They recognize that when professionals are aware of their own values, attitudes, cultural background, and personal needs, the process of helping others is less likely to be negatively impacted by those factors.

Statement 37. Human service professionals recognize a commitment to lifelong learning and continually upgrade knowledge and skills to serve the populations better.

REFERENCES

Southern Regional Education Board. (1967). *Roles and functions for mental health workers: A report of a symposium.* Atlanta, GA: Community Mental Health Worker Project. Used with permission.

Reprinted with permission of *Human Service Education*. Ethical standards of human service professionals. (1996) *Human Service Education 16*(1), 11–17.

Index

Ability(ies), 16, 24, 27, 41, 55,
  57, 59, 64, 82, 92, 93, 104,
  110, 111, 114, 125, 131,
  136, 145, 167, 172, 175,
  187, 215, 219, 220, 233
Academic
  curriculum, 6
  department, 189, 190, 216,
    219, 222
  expectations, 187
  journal articles, 46
  knowledge, 56
  preparation, 191, 236
  program(s), 6, 192, 236
Accountability, 213, 219, 222,
  223
Accreditation, 151
Acquired skills, 2, 235
Action(s), 59, 96, 121, 122,
  132, 142, 143, 173, 174,
  178, 188
Active learner, 66
Active learning, 7
Activity(ies), 4, 8, 26–28, 30,
  38, 51, 67–69, 131, 139,
  142, 144, 146, 152, 156
Acts of violence, 206
Adjustment(s), 9, 26, 181 192
Administration, 42, 58, 72, 74,
  238
Administrative
  assistants, 36
  internship placement,
    151–157
  role(s), 16, 62, 139, 151,
    157
  services, 22, 151–157
  services situation, 71–74
Administrator(s), 22, 37, 73,
  151, 152, 156, 173
Advocacy, 22, 120
Advocate, 21, 43, 82, 120
Age
  difference, 85
  distribution, 49
  group(s), 16, 17, 102
Agency(ies), 2–4, 8, 15, 16, 18,
  24–26, 34–53, 90, 102,
  120–123, 127, 128, 131,
  139, 140, 142, 150, 151,
  155, 160, 164, 168, 169,
  171, 182, 190, 192, 198,
  202, 207, 209, 222, 223,
  227, 233, 237, 238, 243,
  245

administration, 10
administrators, 8, 141
board of directors, 160
brochure, 157
clients, 103
culture, 10, 112–114
director, 35, 125, 152
effectiveness, 153
functions, 41
funding, 42–43, 47, 141, 143
goals, 37–40, 154
identity, 34
mission, 37–40, 47, 146, 236
network, 43–45
newsletter, 156–157
objectives, 37–40
performance, 152
philosophy, 37
policies, 189
policy(ies), 11, 67, 118
strategies, 37–40
worker function, 20–23
African American, 99
Aggressive
  behavior, 167, 168, 206
  incidents, 207
  responses, 170
Alcohol, 164, 183, 190
American Psychological
  Association (APA), 103
American Red Cross, 42
Anger, 168, 168, 172, 207, 230,
  241
  management, 172
Anglo-American, 110
Angry, 172, 194, 202, 209
Anxiety(ies), 8, 36, 83, 159,
  160, 179, 187, 190, 206,
  207, 241
Anxiety-provoking experiences,
  159, 165, 213, 218
Appalachian mountain culture,
  110
Appearance, 107
Application of knowledge, 2,
  59, 62, 67
Appreciation, 172, 235, 244,
  245
Apprenticeship, 3
Articulate learning, 65–66, 67,
  71, 74, 78, 79
Asian, 99
Assertive, 87, 171, 172
  communication, 169–171,
    181, 185, 207

nonverbal behavior, 167
response, 168, 170
Assertiveness, 34, 111,
  167–171, 205, 220
Assessment, 47, 67, 148, 215
Assignment(s), 3, 5, 8, 11, 26,
  67, 125, 142, 202
Assumptions, 20, 56, 60, 61,
  64, 74, 100, 107, 112
Attacks, 169
Attitude(s), 6, 9, 27, 51, 52, 58,
  61, 90, 100, 104, 105, 107,
  124, 127, 179, 181, 182,
  195, 228
  judgmental, 115
  positive, 87
  social, 50
Autonomy, 4, 7, 10, 58, 67,
  106, 112, 131
Awareness, 16, 61, 65, 89, 100,
  105, 106, 136, 166, 198,
  209, 235

Bad habits, 140
Barriers, 85, 115, 177
Behavior(s), 9, 35, 55, 58–61,
  64, 86, 90, 103, 105,
  110–112, 114, 120, 121,
  124–127, 142–143, 160,
  165, 170, 174, 178, 185,
  188, 190, 192, 193, 201,
  206, 207, 213, 215, 228,
  241
  change, 21
  inappropriate, 189
  pattern(s), 11, 203
  violent, 209
Belief(s), 56, 89, 92 100, 101,
  105, 107, 112, 118, 177,
  179–181
Biases, 20, 27, 90, 107
Bicultural(ism), 101, 110–112
Black, 99, 106
Board of directors, 37, 152, 157,
  159
Body language, 159, 161, 177
Boundary(ies), 20, 43, 47, 196,
  204, 206
  issue(s), 193, 200, 206
  violation, 196
Brochure(s), 35, 156, 192
Broker, 21, 82
Budget, 42, 154
  needs, 43
  proposals, 35

Bureaucrats(ic), 43, 141
    organization structure, 40
    pyramid structure, 40
    requirement, 142
Burnout, 73, 74, 201–202,
    203
Business(es), 110, 112, 195,
    199, 227
    deal, 199–200
    relationships, 198
By-laws, agency, 152

Campaign, fund-raising, 195
Campus, 16, 114
    culture, 113, 114
Care plan, 145, 146, 149
Care facilities, residential, 142,
    148
Career(s), 2, 4, 15, 66, 83, 92,
    118, 120, 136, 139, 140,
    164, 171, 172, 175, 178,
    184, 200, 205, 206, 223,
    235
    choice, 220
    goal(s), 5, 24, 168
Case
    conference(s), 4, 114, 157
    record, 142
    statement(s), 153, 154
Census projections, 99
Ceremonial close, 244
Challenge(s) (d), 5, 8, 9, 11, 43,
    85, 99, 100, 104, 111, 142,
    149, 164, 175, 204, 205,
    226, 229, 235, 241
Classroom, 2, 6, 24, 56, 62,
    108, 187, 205
Clerical tasks, 4
Client, 4, 8, 11, 20–22, 30, 35,
    36, 38, 41–44, 50, 51, 53,
    67, 88, 89, 91, 100, 101,
    103, 104, 108, 110–112,
    114, 115, 118, 127, 136,
    139–146, 148–151, 159,
    160, 164, 171, 175, 187,
    189, 192–196, 202, 206,
    219, 226–231, 233, 235,
    238–242, 244–246
    aggressive, 207–208
    agitated, 208
    anxiety, 238
    attitudes, 107, 177
    background, 102
    behavior, 107, 119, 120, 122,
        143
    care, 122, 142, 144, 145, 238
    care plans, 145–148
    chart 150,
    concerns, 143,
    confidentiality, 131
    contact notes, 141, 148–149
    culture, 64, 103
    dignity, 64
    feelings, 177
    group(s), 22, 45–47

growth, 231
history, 142
needs, 142, 144, 145, 177,
    223
population(s), 24, 46
privacy, 122
record(s), 4, 139, 140, 143,
    144, 145, 149, 151, 207
resistance, 240
respect, 119
right(s), 21, 120, 131
self-determination, 64, 123
strengths, 120
traditional groups, 238
values, 177
violent, 121
well-being, 120, 141
Closing summary (ies),
    150–151, 226
Closure, 7, 227, 242, 244
Clothes, 108
Co-worker(s), 28, 36–37, 55,
    160, 172, 178, 180, 182,
    192, 195, 196, 199, 201,
    226, 227, 229, 244
Coerciveness, 207
Cognitive-behavioral approach,
    174, 175
Collaborative methods, 40
Colleagues, 8, 11, 36, 90, 120,
    121, 134, 135, 141, 196
College, 1, 113, 190, 192, 197,
    222, 228
    classes, 114
    employee, 222
    library, 46, 47
College-age young adults, 17
Comfort, 4, 10, 16, 25, 64, 86,
    244
Commitment(s), 120, 190, 192,
    195, 201, 213, 223
Communicate, 16, 86, 152,
    167, 172, 181, 201, 205,
    244
Communication, 40, 83, 85, 96,
    106, 127, 159, 167, 171,
    172
    aggressive, 169, 171, 172
    internal, 151
    intra-agency, 151
    open, 9
    patterns, 100
    skill, 159
    two-way, 86, 220
Community, 35, 38, 40, 44,
    47–51, 71, 82, 111, 112,
    113, 120, 128, 154, 156,
    213
    awareness, 156
    education, 20, 125
    group(s), 21, 22, 37, 47, 157,
        158
    organizations, 42, 151
    planner, 22
    problem, 43

Community-based organiza-
    tions, 47
Competence, 5, 7, 100, 105,
    114, 223
Competent, 11, 174
    practitioners, 223
Competition, 43, 45
Compromise, 198
Computer
    files, 151
    skills, 192, 220
Concern(s), 18, 19, 25, 41, 72,
    83, 84, 89, 108, 128, 143,
    144, 146, 160, 168, 169,
    182, 183, 193, 198, 205,
    213, 235, 243
Confidence, 10, 25, 64, 233
Confident, 167, 175, 187, 214
Confidential(ity), 5, 64, 119,
    120, 121, 122, 131, 182
Conflict(s), 43, 45, 64, 73, 74,
    93, 110, 118–121, 125,
    131, 169, 171–174, 188,
    193, 197, 205, 218
    data, 63
    expectations, 112
    interpersonal, 166, 171, 180
    relationship, 179
    resolution skills, 171–174,
        181, 185
Confrontation, 7, 126
Consensus, 51, 123
Constituencies, 219
Consultation(s), 64, 120, 133,
    135, 136, 157
Consultant, 22, 174
Consumers, 213
Contact(s), 15, 89, 91, 96, 148,
    149, 238, 240, 244
Continuity, 239
    of care, 142, 242
    of services, 142
Contract(s), 42, 190, 215
Control, 9, 70, 72, 73, 74, 125,
    165, 172, 180, 181, 228
Cooperation, 43, 120
Coordinator(s), 40, 125
    program, 125, 151
Cope(ing), 100, 101, 110
    ability, 210, 241
    skills, 205
    strategy(ies), 181
Corporate culture, 111, 112
Council on Standards for
    Human Services Education,
    2
Counseling, 16, 21, 90, 202,
    205
Counter-productive, 110, 165,
    182–183, 193
    language habits, 160
Course(s), 107, 149, 150, 152,
    160, 184, 188, 222, 227,
    228, 233, 234
    goals, 216

objectives, 26, 215
  of action, 67, 131, 133,
    135–136
  syllabus, 5, 215
Coursework, 1, 56, 124, 227
Crisis, 141, 164, 205
Crisis intervention agencies, 42
Crisis services, 19, 51
Critical thinking, 2, 3, 80
Criticism, constructive, 180
Critique, 3, 153, 154, 155, 174
Cross-cultural, 100, 114
  competence(ies), 100, 107
  context, 111
  helper(s), 100, 101, 105
  helping relationship, 104
Cultural
  assets, 104
  awareness, 106
  background(s), 100, 104,
    105, 107
  barriers, 104
  bias, 115
  challenges, 114
  characteristics, 100
  competence, 104, 115
  context, 103
  contrast(s), 105, 106
  differences, 104, 115, 120
  diverse populations, 120
  diversity, 49, 115
  group(s), 99, 100, 102, 104,
    105, 106
  heritage, 106, 112
  identity, 102, 110, 111, 112
  influences, 109
  insight, 108
  learned behavior(s), 103,
    105, 110, 111, 112
  lens(es), 105, 108
  make-up, 101, 105, 106, 108,
    115
  norms, 112
  patterns, 107, 111
  self-awareness, 115
  strengths, 104
  traits, 102, 107
  value, 110
  variable, 106
Cultural bias, 109
Culture(s), 99, 100–107, 109,
    112, 115, 120, 244
  clashes, 110, 112, 114
  different, 104, 107, 110
  dominant, 110, 111
  primary, 110, 111
  shock, 104

Danger, 148, 189
Data manager, 22
Decision making, 7, 15, 50, 67,
    111, 118, 119, 127,
    130–136
Depression, 175, 206
Developing a plan, 67–68

Development, personal, 27, 28,
    29, 31
Dilemma(s), ethical, 118, 133,
    134
Direct service(s), 20, 22, 58, 62,
    74
  care, 21
  clients, 142–151
  role(s), 139, 157
  situation, 69–71
  work, 238
Director, 72, 190, 202
  program, 40, 151
Disabilities, developmental, 19,
    141
Disagreement, 64, 172
Disappointment(s), 3, 199
Disclosure, personal, 240
Discomfort, 182
Discouragement, 51
Discussion(s), 15, 88, 90, 91,
    96, 108, 135, 183, 190,
    192, 198, 219–221, 231,
    236, 240
Disengage, 229
Dissonance, 58, 63–64, 70, 73,
    78, 125
Diversity, 99–116
Document(s), 26, 30, 128, 139,
    142–145, 153, 154, 215
Documentation, 122, 140,
    141–145, 150, 215
Donations, 154
Donors, 153, 154
Dress, 24, 25, 112, 189
Dual
  mission, 42
  purposes, 96
  relationships, 196–200

Elderly, 16, 44, 51, 69
Electronic sources, 47
Emotional(ly), 171, 172, 174,
    185, 202, 242
  challenging, 164,
  closure, 231
  displays, 160
  distress, 175
  exhaustion, 206
  intensity, 11, 226, 240
  overburdened, 206
  reaction(s), 164, 165, 166,
    179, 185, 228, 240, 241,
    243
  support, 229, 240
  work, 227
Emotions, 1, 8, 34, 55, 58, 60,
    71, 88, 120, 164–168, 171,
    174, 175, 177, 179, 185,
    227, 230, 246
Empathy, 123, 164, 240, 241
Employee(s), 1, 35, 36, 38, 74,
    112, 171, 189, 190, 192,
    193, 230, 238
Employer(s), 120, 146, 222

Employment, 3, 16, 146, 223,
    236
Enabler, 82
Environment(s), 5, 21, 25, 41,
    47, 50, 70, 73, 100,
    108–110, 112, 121,
    179–182, 191
Erikson, Eric, 7, 83
Ethical
  action, 119, 123, 133
  code, 119
  competence, 118–137
  constraints, 121
  decision-making, 118, 133,
    136
  dilemma(s), 64, 118, 119,
    131
  foundation, 118–121
  guidelines, 133, 134, 196
  issues, 118–137
  knowledge, 125
  positions, 119, 130–136
  practice, 120
  principle(s), 64, 66, 130, 131,
    133, 134
  principles screen, 130–133,
    136
  problems, 198
  responsibilities, 118
  risk factors, 123
  scandals, 118
  sensitivity, 125
  skills, 118
  standard(s), 118–123, 125,
    136
  values, 119, 134
  violations, 120, 124, 125
Ethical Standards of Human
    Service Professionals, 119,
    123
Ethics, 90, 187
Ethnic
  community, 111
  differences, 100
  group(s), 49, 100, 101, 111
Ethnocentric(ism), 104,
    107–110, 115
  reaction(s), 107, 108, 109,
    115
Evaluation, 22, 60, 129, 141,
    192, 193, 212–225, 227,
    235, 236, 242, 246
  conference, 219–222
  feedback, 212
  final, 218
  formal performance, 212,
    227, 236
  instrument, 5
  of practitioners, 213
  open-ended, 218
  preparing for, 215–218
  process, 11, 152, 212, 213,
    214, 215, 218, 219, 223
  program, 152
  standards, 218

Expectations, 2, 4, 5, 6, 9, 12, 24, 83–85, 88, 112–114, 121, 123, 179–181, 187, 190, 192, 199, 213, 214, 216, 223, 237
Expertise, 18, 22, 83, 86, 120, 154
Exploitation, 71, 122, 198
Eye contact, 24, 159, 160, 167, 207

Faculty, 2, 23, 216
    liaison, 4, 12, 24, 30, 46, 47, 85, 182, 190–193, 195, 202, 213, 218, 222–223, 242
    member(s), 6, 7, 8, 15, 16, 20, 23, 26, 127, 222, 223
    perspective, 1
    supervisor, 58, 127, 174, 222
Failure, 175, 208
Fear(s), 18, 83, 110, 179, 208, 241
Federal
    funds, 42
    government, 51
    laws, 120
    policies, 52
Feedback, 44, 67, 87, 88, 92–96, 160, 188, 212, 219, 221, 223, 237
    negative, 92, 216
    positive, 92
Feeling(s), 20, 56, 61, 63, 72, 74, 83, 90, 104, 107, 108, 125, 160, 164–186, 188, 206–208, 219, 227–231, 236, 240, 241, 244, 245
Field, 1, 5, 18, 22, 26, 55, 62, 72, 118, 120, 213, 220, 222
    agency(ies), 46, 55, 108, 112, 119, 134, 139–163, 166, 170, 193, 194, 222
    experience(s), 2, 3, 4, 6, 26, 45, 58, 62, 64, 92, 99, 101, 103, 173, 198, 214, 216
    instructor(s), 30, 32, 86, 87
    interests, 15
    of practice, 16, 18–20, 23
    placement(s), 3, 5, 8, 23, 24, 58, 88, 108, 113, 127, 129, 170, 173, 188, 216, 218, 223, 227, 229, 235, 236, 238, 242
    programs, 227
    seminar, 30, 58, 128, 182
    site selection, 16
    site(s), 4, 5, 9, 15–16, 18, 20, 23, 32, 34, 35, 37, 44, 46, 50, 236
    supervision, 218–219
    supervisor(s), 4, 8, 11, 12, 23, 24, 26, 27, 30, 32, 35, 46, 47, 56, 58, 67, 82–86, 88, 95, 108, 125, 167, 189,
190, 192, 195, 212, 213, 217, 218, 219, 223, 231, 236, 237, 242
    evaluation, 218–222
Fieldwork, 2, 3, 5–8, 12, 20, 22, 23, 32, 37, 46, 47, 53, 55, 57–59, 62, 65, 67, 80, 93, 92, 95, 101, 105, 106, 108, 109, 118, 119, 124, 127, 132, 142, 158, 160, 161, 165, 167, 174, 180, 182, 197, 212, 214, 222, 223, 226, 227, 241, 242
    learning plan, 26–28
    requirement, 6
Financial
    accountability, 22
    information, 152
Flexibility, 74, 85, 108, 173, 192, 200, 220
Flight into illness, 240–241
For-profit organizations, 42, 43
Frustration(s), 17, 74, 110, 164, 168, 178, 199, 230
Full disclosure, 131
Fund-raising, 38, 43, 153, 154
    efforts, 20, 154
    presentations, 154
Funding, 37, 42–43, 51, 53, 143, 154, 155
    method(s), 43
    requests, 153–156
    source(s), 42, 143, 151, 156, 157
    tool 154
Funds, 153, 154

Gatekeeper, 223
Geographic areas, 47, 49
Gestalt idea, 245
Gifts, 244
Goal(s), 16, 20, 25–28, 30, 32, 37–40, 43, 82, 85, 90, 142, 145–147, 149–152, 165, 201, 215, 216, 218–221, 223, 232, 235
    of supervision, 198
    setting, 7, 27
    statement, 27
Goal-setting process, 27
Going-away, 244, 245
Good-bye, 227, 228, 236–242, 244, 245, 246
Government contracts, 42
    documents, 47
    funds, 42
    policies, 36
    programs, 42
    purchase of services, 42
    representatives, 153
    sources of funding, 42
Grade(s) in fieldwork, 5, 218, 222, 223
Grammar, 87, 145, 191
Grant(s), 42, 155, 156, 233

proposals, 35, 153, 154, 155
    sources, 153
    writing, 154
Group(s), 16, 17, 21, 45, 72, 74, 99, 102–104, 106, 111, 112, 127, 139, 159–161, 244, 245
    cooperation, 38
    diverse, 99, 100–102, 109
    dominant, 100
    social, 43
    socioeconomic, 101, 111
Group treatment settings, 239
Growth, 7, 43, 83, 88, 219, 221, 223, 226, 233, 240
    personal, 4, 7, 109, 120
Guidelines, 50, 93, 94, 95, 142, 159, 160, 196, 227–235
Guiding principles, 136, 144, 235, 240
    documentation, 144–145
Guilt, 204, 226, 241

Habit(s), 7, 27, 58, 64, 66, 110, 115, 140, 160, 183, 201
"Halo" effect, 215
Harm, 131, 195, 199
Healthy, 164, 178, 184
    attitudes, 182
    balance, 201, 204
    boundaries, 209
    relaxation, 183
Hierarchy, 130, 131, 132
    of ethical principles, 136
High risk cases, 207
Hispanic people, 99, 112
History(ies), 71, 102, 147, 206, 219, 240
    personal, 60
    of problem, 146
    social, 146, 147
Home visits, 148, 207, 208
Human service(s), 2, 3, 6, 36, 72, 82, 92, 104, 119, 120, 127, 139, 159, 164, 196, 227
    agencies, 16, 18, 20, 22, 37, 42, 43, 89, 114, 129, 142, 154, 156, 171, 172, 187, 213, 244
    classes, 139
    delivery system, 50
    department, 8, 202
    education, 61, 223
    educators, 213
    evaluations, 213
    experience, 235
    faculty objective, 219
    intern(s), 86, 87, 125, 141, 178, 182, 193
    internship, 164, 187
    intervention, 46, 144
    job, 234
    literature, 133, 177, 198
    major, 192

organization(s), 36, 53, 139, 145, 156, 171, 192
practice, 119, 123
profession, 118, 119, 123, 223
professional(s), 3, 6, 25, 107, 118, 120, 123, 125, 161, 175, 187, 200, 206, 223
professional responsibilities to clients, 119–120
program(s), 5, 15, 20, 22, 26, 27, 118, 187, 215
settings, 119, 122, 139, 140, 141, 157
values, 123
work, 3, 118, 120, 178, 242
worker(s), 20, 50, 51, 55, 83, 99, 100, 103, 107, 109, 110, 112, 115, 120, 121, 123, 129, 130, 140, 141, 142, 167, 171, 193, 201, 205, 206, 213
worker's roles, 161

I-statements, 167, 169
Identity, 7, 37, 43, 53, 106
Indirect services, 58, 71
Initiation stage of development, 8–9
Integrative processing model, 55–81, 125, 166
Intervention(s), 20, 21, 59, 64, 88, 122, 130, 131, 142, 143, 148, 149, 171
Interview, 8, 24, 28, 35, 39, 71, 146, 148, 149, 159, 234
Italian-American, 111

Job(s), 5, 15, 16, 34, 36, 72, 73, 83, 110, 140, 145, 188, 189, 192, 200, 206, 214, 216, 218, 219, 222, 227, 230
interview, 3, 24, 222, 233
opportunity, 156
search, 16, 36
title, 37, 40
Journal(s), 30, 57, 58, 80, 166, 177
Journal articles, 103

Knowledge, 2, 5, 6, 10, 18, 22, 27, 28, 37, 45, 46, 55–58, 62–67, 73, 82, 83, 85–87, 90, 100, 102, 103, 115, 119–124, 160, 180, 219, 235, 241
base, 47
of content, 58
of theory(ies), 56, 58, 62
relevant, 58, 62, 70, 77
theoretical, 66, 118, 191

Language, 37, 100, 106
judgmental, 144, 160

patterns, 160
slang, 160
subjective, 144, 160
Latino people, 102
Law(s), 50, 52, 120, 128, 133, 135
Leadership, 26, 73, 74
functions, 20
theory, 62
Learn(ing), 2, 3, 6, 16, 18, 24, 26, 27, 30, 34, 35, 55, 58–60, 63–67, 71, 72, 80, 82, 83, 85–87, 91, 92, 96, 103, 105, 107, 164, 165, 168, 175, 179, 181, 188, 193, 198, 200, 202, 205, 206, 216, 219, 223, 227, 232, 233, 234, 235
agency procedures, 10
agreement/contract, 9, 26, 215, 216
cycle, 67
from experience, 2, 23, 55–68, 64, 220, 237, 55–68
goals, 1, 9, 10, 26
needs, 8, 9, 10, 15, 23, 32, 85, 219, 235, 236
plan, 15, 26–28, 30, 32, 89, 215
process, 26, 59, 67–68, 87
Leave(ing), 189, 207, 229, 238, 244
Legal
constraints, 4, 121
contexts, 50–52
documentation, 142
environment, 47
evidence, 142
foundation(s), 119, 121–123
issues, 28, 30, 47, 53, 118, 119, 121–123, 124–130, 133
liability, 118
obligations, 121
process, 30
purposes, 27
questions, 122
rights, 28, 30
risk factors, 123
standards, 119, 121–123, 124, 136
Limit setting, 111
Limitations, 8, 48, 120, 181
Listen(ing), 72, 160, 172, 183, 207, 220
Listener(s), 87, 160
accepting, 231
active, 166
good, 231
Litigation, 118, 122
Loewenberg and Dolgoff, 130

Malpractice, 123
Managed care plans, 51

Management, 74
nonhierarchical philosophies, 40
Managers, midlevel, 40
Maslow, 109
Measurement methods, 30
Meeting(s), 3, 4, 8, 23, 24, 26, 28, 32, 73, 74, 83, 88, 89, 136, 152, 159, 166, 207, 242
initial, 15, 23–25, 26, 88, 154
Mental habits, 175
Mentor, supportive, 82
Mentoring, informal, 36
Mission statements, agency, 37–40
Mistakes, 175, 187, 188–191
Motivation, 9, 18, 20, 73, 142, 161
Myths, 1, 2–5

National Organization for Human Service Education, 196
Native American, 99
Network(ing), 3, 36, 43–45
News magazines, 51
Non-Western cultures, 111
Nonassertive behavior, 168, 169, 171
Nonverbal communication, 59, 60–61, 100, 106, 159, 161, 167, 241
Norm(s), 88, 100, 111, 112, 114, 146
Not-for-profit organizations, 43
Note(s), 148, 160, 215
progress, 148–149

Objectives, agency, 37–40
Observation(s), 2, 7, 8, 9, 34, 35, 40, 41, 47, 59, 108, 125, 128, 143, 178, 205, 234
Observe(ing), 3, 4, 28, 30, 41, 59, 88, 101, 105, 108, 113, 175, 176
Observer, 35, 41, 59, 87, 119
Obstacle(s), 57, 64, 67, 71, 85, 104, 110, 216
Oral communication presentation skills, 158, 160, 161
reporting, 157–161
reports, 139, 157
Organization(s), 2, 3, 4, 8, 9, 20, 23, 24, 34, 35, 36, 37, 38, 39, 40, 41, 42, 44, 47, 50, 55, 72, 82, 87, 102, 110, 112, 123, 136, 145, 151, 152, 154, 156, 157, 171, 172, 192, 219, 227, 242, 244
formal, 41

informal, 40, 41
national, 42
proprietary, 42
seeking funds, 153
structure, 40–41, 148
theory, 62
vertical chart, 40
Orientation, 7, 8, 34, 104
Outcome(s), 133, 146, 169, 170, 172, 173, 215
Outreach, 21
Overdependency, 241

Paperwork, 136, 139, 140, 141, 151
Participant-observer, 59
Performance, 55, 141, 175, 184, 187, 192, 205, 206, 212–225
evaluation, 141
improvement, 190
quality, 212
Philosophy(ies), 38, 40, 43, 52, 190
Physical aggression, 206
appearance, 35, 146
characteristics, 35, 100
exhaustion, 206
harm, 189
safety, 206
stance, 208
Placement(s), field, 2, 4, 5, 8, 16, 23–26, 35, 56, 113, 126, 127, 132, 141, 188, 192, 199, 219, 222, 229, 231, 232, 236, 238, 242
Policy(ies), 4, 38, 50, 51, 188, 189
changes, 152
manual, 35
operating, 207
social, 36, 50, 51
Policymaking, 52
Political issues, 47, 50–52, 53
Population(s), 3, 16–17, 20–23, 42, 45, 46, 52, 99, 100, 101, 103, 121, 122, 206, 207, 209, 238
Position, entry-level, 192
Posture, 167
nondefensive, 165, 220, 221
open, 165, 221
Power, 86, 100, 130, 173, 199
Practice, 2, 6, 27, 45, 50, 56, 58, 59, 64, 67, 75, 81, 101, 119, 125, 136, 140, 151, 157–159, 161, 167, 171, 172, 177, 201, 207
situation(s), 119, 124–130, 236
Practitioner(s), 47, 213, 219, 223
Pre-placement stage of development, 8
Premature disengagement, 11, 229

Presentation(s), 30, 72, 73, 74, 159, 160
Press releases, 156
Prevent(ion), 20, 136, 187, 207, 208
Privacy issues, 5, 119, 131
Private agencies, 42
contributions, 42
donations, 42
donors, 153
for-profit agency(ies), 42
foundations, 42
grants, 143
nonprofit organizations, 42, 153
privileges, 4
Problem-solving, 167, 168, 169, 170, 172
Profession(s)(al), 3, 22, 36, 44, 56, 56, 61, 63, 64, 72, 74, 82, 112, 118–123, 136, 139, 140, 145, 151, 157, 158, 164, 172, 174, 175, 196, 202, 206, 213, 215, 219, 223, 236
books, 103
boundaries, 193–196
development, 11, 22, 36, 223
dress, 4
ethics, 58, 118, 119, 122, 124, 187
growth, 4, 120, 218, 223, 235
helping relationship, 246
identity, 18
issues, 88, 90, 96, 127
journal articles, 46
literature, 45
manner, 165
meeting, 142
misconduct, 123
relationships, 200
responsibility, 10
role, 142, 193, 196
role models, 6, 127
skill(s), 4, 83, 136, 213, 236
values, 90, 123, 124
wisdom, 208
Professor, human services, 3, 26
Project(s), 140, 151, 154, 167, 227, 242, 243
administrative, 192, 242
goal, 154
special, 157
Public, 153
accountability, 42
agencies, 42, 43
nonprofit organizations, 42
opinion, 50
relations, 156–157
revenues, 143
safety, 122

Quality
of agency work, 143
of care, 142

of life, 131
of services, 120, 143

Rating error, 215–217
Read(ing), 8, 28, 46, 62, 67, 104, 140, 149, 151, 160, 183, 201, 227
list, 46, 103
records, 146
Reassurance, false, 11, 193, 229
Records, secure, 120
Record-keeping, 140
Referrals, 44, 120, 141
Reflection(s), 3, 20, 56, 125, 166, 192, 223, 227, 235, 240, 246
Regret(s), 227, 229, 241, 245
Relaxation, 159, 165, 183
Relaxation-inducing activities, 181, 185
Reliability, 4, 87, 122
Reparable mistake, 189
Report(s), 40, 122, 128, 139, 142, 144–146, 148–151, 159, 208, 240
annual, 35
interim, 152
mock, 140
periodic, 35, 152
progress, 152
to board of directors, 152
Reporting, 139, 153, 161
Research, 152, 206
findings, 118
interests, 47
papers, 3
skills, 47
Responsibility(ies), 4, 7–11, 25, 26, 30, 55, 56, 68, 87, 90, 103, 106, 136, 140, 145, 150, 164, 165, 180, 185, 191, 192, 200, 202, 205, 213, 219, 222, 223, 239, 243
Resume, 3, 24
Review(s), 133, 149, 150, 171, 213, 231–233
periodic, 149–150
team, 157
Rituals, 34, 112, 244, 245
Role(s), 2, 5, 20, 73, 83, 96, 108, 128, 139, 196, 198, 218, 222, 223, 237, 241
functional, 20, 22, 23
good model, 167
play, 67, 71
social, 206

Sadness, 227, 230, 240, 241
Safe, 206–209
Safety, 120
measures, 207
policies, 189
regulations, 189
Schedule, 201, 202

arrangements, 2
meetings, 157
Secretaries, 36
Security, 207
Self-actualization, 109
Self-assessment, global, 216
Self-awareness, 9, 58, 61, 109,
120, 123, 165, 166,
184–185, 228
Self-care, 70, 204, 206
Self-control, 177–179, 185
Self-discipline, 200
Self-disclosure, 240
Self-esteem, 174
Self-evaluation, 161, 218
Self-observation, 166
Self-restraint, 178
Self-talk, 175, 177
negative, 175, 176, 177
positive, 174–177, 181, 185
realistic, 181
Self-understanding, 166, 185
Seminar(s), 3, 67, 182
Separation(s), 227–229
Site visit, 127
Situation, high-risk, 71, 207,
208
Skill(s), 2, 3, 5, 6, 9, 11, 16, 18,
20, 27, 29, 31, 55, 59, 64,
67, 68, 71, 72, 80, 82, 87,
90, 91, 99–101, 114, 119,
120, 140, 154, 157,
159–161, 164, 171, 172,
181, 184, 206, 216, 219,
222, 235
SOAP approach, 148–149
Social agencies, proprietary, 42
Social problems, 48, 50, 51, 52,
139
Social policy, formal, 50
Southern Regional Education
Board, 20
Staff, 9, 35, 37, 53, 56, 121,
123, 128, 136, 145, 157,
158, 178, 194, 195, 198,
206, 226, 238
cohesion, 112
meeting(s), 4, 205
staff member(s), 4, 28, 34,
36, 37, 41, 44, 56, 66, 89,
91, 112, 118, 121, 127,
128, 131, 140, 143, 151,
157, 158, 177, 187, 190,
195, 199, 209, 242
morale, 47
training, 38, 207
workloads, 51
Staffing(s), 157, 160, 197, 207
Stage(s) of internship developmental, 8–12, 83
Standards, 121, 190, 222
of care, 122,
of competence, 213
of practice, 123
of the profession, 121–123
Stereotype(s), 20, 56, 102, 107

Strategy(ies), 146, 154, 164,
172, 173, 181, 182, 216,
228, 235
Stress(es), 47, 51, 72, 83, 110,
164–186, 200, 203, 231
boundaries, 204
management, 72, 179–184
reactions, 175
reduction, 183
reliever, 182
response, 180, 181, 183
Stressors, 171, 179, 181, 182,
183
external, 180, 181, 182
internal, 180, 182
Student-supervisor relationship,
236
Subgoals, 145
Summary
note(s), 148, 149
transfer, 226
Summative evaluation, 212,
223
Supervision, 2, 10, 11, 38, 58,
64, 82–98, 108, 160, 192,
199, 202
feedback, 92–96
Supervisor(s), 3, 4, 8–11, 15,
23–26, 30, 32, 36, 38, 40,
41, 44, 56, 58, 59, 66, 67,
71, 82, 83, 85, 88, 90, 94,
95, 108, 114, 123, 139,
141, 143, 151, 157,
159–161, 166, 174, 178,
180, 181, 187, 188,
190–193, 195, 197–199,
201, 202, 204, 205, 207,
212, 213, 218, 221, 226,
227, 236–239, 242, 244,
245
assessment, 219
attention, 89
characteristics, 84
evaluation, 11, 141, 219, 220
expectations, 83, 88
feedback, 140, 219, 220, 221
permission, 189
Supervisory, 199
conference(s), 67, 90, 195
expectations, 191
process, 86
relations, 94–95
relationship(s), 82–86,
88–96, 198, 199, 200,
236
sessions, 67, 89
time, 89, 91, 96

Tardiness, 189, 192
Target group, 16, 43, 46, 20
Teacher/educator, 21
Tension, 43, 47, 179, 199
Terminate, 150 , 151, 229, 246
Termination, 189–191,
226–247
patterns, 227–229

positive, 11, 226–247
process, 226–247
rituals, 244–245
stage of development, 11–12,
226–247
Textbooks, 56
Thank-you note(s), 25, 244
Time-management skills, 205
Transfer summaries, 242
Transference, 241
Trouble-shooting, 187–211
Trust, 9, 82, 83, 208, 218, 240

United State Bureau of Census,
47, 99
Unprofessional, 165, 189

Value(s), 6, 11, 23, 26, 27, 35,
37, 38, 40, 50, 51, 52, 58,
60, 61, 64, 70, 90, 100,
104, 107, 110, 112, 119,
120, 123–124, 136, 140,
174, 187, 220, 231
issues, 119, 124–130
personal, 64, 118
positions, 130–136
systems of, 112
trait, 145
Verbal communication, 59, 60,
100, 106, 159, 160, 167,
207
Violence, 206–207
Vocal expression, 159, 161, 177
Voice, 123, 159, 167
Volunteer(s), 1, 4, 16, 26, 66,
154, 230

Weaknesses, 9, 24, 48, 93, 133,
152, 157, 212, 213, 216,
218, 219, 223
Win-lose negotiations, 172
Win-win conflict resolution,
172
Withdrawing from fieldwork,
229
Work, ethic, 111
habits, 141, 187, 191–193,
201, 203, 209
life, 183, 200–206
responsibilities, 200, 229
schedules, 82
transferring, 242–244
Work-related stress(es), 180,
182
Worker(s), 11, 18, 36, 57, 64,
67, 91, 100, 107, 119–123,
130, 131, 139, 141–146,
148–151, 158, 161, 171,
178, 188, 191, 194–196,
207, 208, 238, 240–242,
245
Worker-client relationships,
198, 239
dual, 120
Working, 83, 164, 183, 201,
207, 219, 227

conditions, 72, 120
relationship, 83, 88, 227
stage of development, 9–10
Workload, 47, 181, 233
Worldviews, 103
Write, 144, 216, 219, 242
Writing, 123, 139–141, 144,
    145, 150, 151, 156, 161,
    166, 192, 226, 233, 244
  administrative, 153

process, 59
responsibilities, 140, 142
samples, 140
skills, 139, 140, 220
Written, 150, 152
  assignments, 222
  communication, 10, 160
  documentation, 139–157
  evaluation, 219
  form, 39

goals, 38
materials, 154, 156, 157
mission statement, 37
record, 142
report(s), 69, 139–157
reporting, 139–157
summaries, 67, 89
work, 140, 142